A Catalogue
of the
Flora of Arizona

A Catalogue of the Flora of Arizona

J. Harry Lehr

Curator of the Herbarium
Desert Botanical Garden

DESERT BOTANICAL GARDEN
PHOENIX, ARIZONA 1978

INTRODUCTION

In 1960, an "Arizona Flora" with a supplement (1) was published and in 1964 a second printing of it appeared, thus, in reality eighteen years have elapsed since the last presentation of the entire flora of the State. In that interval extensive floristic work and exploration have been accomplished and many revisions have appeared for groups containing species in the State. These additions and corrections have now become so numerous that the last edition of the State flora is inadequate to meet the present needs of the botanical community in Arizona. The publication of a flora of North America or a southwestern flora would probably have met this need, but the first, after being initiated, foundered, and the second remains in the discussion stage. This catalogue with its annotations is now presented to satisfy the need for a current checklist of Arizona's flora. In addition, for those who already know the plants, it will be a more handy reference and its size will make it more useful in the field.

This is not a critical taxonomic study and the changes herein are only those that have been generally accepted by the taxonomic community and included in the most recent manuals of the general area. The sequence of families substantially agrees with that in the "Arizona Flora" but in those instances where genera inclusion within particular families has been changed, this change is in agreement with Willis' Dictionary (2). The arrangement of genera within a family, and species within a genus, are strictly alphabetical and when a change in synonymy occurs the only synonym included in this catalogue is the most recent one in the "Arizona Flora." The terms "variety" and "subspecies" are again used with the same lack of uniformity as expressed in the introduction to that flora and the questionable occurrence of a species in this State is again indicated by (?). In an effort to attain some order out of the confusion that exists in the family Cactaceae, I have substituted Benson's "Cacti of Arizona" (3) in its entirety. The extensive synonymy in this monograph is a commendation in itself.

In presenting this catalogue an effort has been made to record the common names under which the plants are known by the layman. In every instance the common name presented here was not manufactured by the writer but occurs either in the cited literature or in the following publications: Ferns: Weatherby (4); Grasses: Hitchcock (5); Trees: Little (6); Weeds: Parker (7); Desert Wild Flowers: Jaeger (8); and the Catalogue of the Flora of Texas: Cory and Parks (9).

An index to families and genera is appended to this study. It is not only cross-indexed but an asterisk precedes those genera which include a new addition to the flora of the State.

The author is grateful to Dr. Charles T. Mason, Jr. of the University of Arizona and Mr. Timothy Reeves of Arizona State University for their suggestions. A particular sense of gratitude is extended to Dr. Donald J. Pinkava of Arizona State University for his generous assistance in reviewing the first draft of this study and his guidance through the complexities of the Compositae. The illustration of *Prosopis pubescens* (Screwbean Mesquite) on the cover was drawn by Wendy Hodgson.

In conclusion, a catalogue such as this, no matter how elaborate, soon becomes outdated and inadequate. Therefore, in order to preserve the current status of this study, the author plans to publish a yearly supplement of additions and corrections in the Journal of the Arizona-Nevada Academy of Science. This will be continued until such time as a revised edition of the "Arizona Flora" or a "Southwestern Flora" is presented to the botanists of this State and its contiguous States.

The
Catalogue

PTERIDOPHYTA. Ferns and Fern Allies.

SELAGINELLACEAE. Selaginella Family.

SELAGINELLA. Spike Moss.

Selaginella arizonica Maxon.
*S. *densa* Rydb. (10).
S. *densa* Rydb.
    var. *scopulorum* (Maxon) Tryon.
S. *eremophila* Maxon. Desert Selaginella.
S. *mutica* D.C. Eaton.
S. *mutica* D.C. Eaton
    var. *limitanea* Weatherby.
S. *x neomexicana* Maxon (S. *mutica x rupincola*).
S. *rupincola* Underw.
S. *underwoodii* Hieron.
*S. *watsoni* Underw. Alpine Selaginella, Watson's Selaginella (10).

PSILOTACEAE. Psilotum Family.

PSILOTUM.

Psilotum nudum (L.) Beauv. Bush Moss, Whisk Fern (11).

ISOETACEAE. Quillwort Family.

ISOËTES. Quillwort.

Isoëtes bolanderi Engelm. Bolander's Quillwort.
I. *bolanderi* Engelm.
    var. *pygmaea* (Engelm.) Clute.

1

EQUISETACEAE. Horsetail Family.

EQUISETUM. Horsetail, Scouring Rush.

Equisetum arvense L. Common Horsetail.

E. arvense L.
  forma ramulosum (Rupr.) Klinge.

E. hiemale L.
  var. *affine* (Engelm.) A. A. Eaton. Scouring Rush, Canuela.

Equisetum laevigatum A. Braun. Smooth Scouring-rush.

E. x ferrisii Clute (*E. hiemale* var. *affine x E. laevigatum*).

OPHIOGLOSSACEAE. Adder's Tongue Family.

BOTRYCHIUM. Grape Fern, Moonwort (12).

Botrychium boreale Milde. Northern Grape Fern.

B. dissectum Spreng.
  forma obliquum (Muhl.) Fern. Cutleaf Grape Fern.

B. dusenii (Christ) Alst. Amphitropical Moonwort.

B. lanceolatum (S. G. Gmel.) Angstr. Lance-leaved Grape Fern.

B. lunaria (L.) Swartz. Moonwort.

B. multifidum (S. G. Gmel.) Rupr.
  var. *intermedium* (D.C. Eaton) Farwell. Leathery Grape Fern.

B. virginianum (L.) Swartz. Rattlesnake Fern.

OPHIOGLOSSUM. Adder's Tongue.

Ophioglossum engelmannii Prantl. Englemann's Adder's Tongue,
  Limestone Adder's Tongue.

O. vulgatum L. Common Adder's Tongue.

SALVINIACEAE. Water Fern Family.

AZOLLA. Water Fern, Mosquito Fern.

Azolla filiculoides Lam. Duckweed Fern.

MARSILEACEAE. Pepperwort Family.

MARSILEA. Pepperwort, Water Clover.

Marsilea vestita Hook. & Grev. (*M. mucronata* A. Br.).
  Hairy Pepperwort, Clover Fern (14).

POLYPODIACEAE. Fern Family.

ADIANTUM. Maidenhair Fern.

Adiantum capillus-veneris L. Southern Maidenhair, Culantrillo.
A. pedatum L. Common Maidenhair.

ASPLENIUM. Spleenwort.

Asplenium adiantum-nigrum L. Black Maidenhair Spleenwort.
A. dalhousiae Hook.
A. exiguum Bedd.
A. monanthes L.
A. palmeri Maxon. Palmer's Spleenwort.
A. resiliens Kunze. Small Spleenwort, Little Ebony Spleenwort.
A. septentrionale (L.) Hoffm. Forked Spleenwort.
A. trichomanes L. Maidenhair Spleenwort.

ATHYRIUM. Lady Fern.

Athyrium filix-femina (L.) Roth
    var. *californicum* Butters. Southern Lady Fern.

BOMMERIA.

Bommeria hispida (Mett.) Underw. Hairy Bommeria.

CHEILANTHES. Lip Fern.

Cheilanthes alabamensis (Buckl.) Kunze. Alabama Lip Fern.
C. covillei Maxon. Coville's Lip Fern.
C. eatoni Baker. Eaton's Lip Fern.
C. eatoni Baker
    forma *castanea* (Maxon) Correll.
C. feei Moore. Slender Lip Fern.
C. fendleri Hook. Fendler's Lip Fern.
C. lendigera (Cav.) Swartz.
C. lindheimeri Hook. Lindheimer's Lip Fern, Fairy Swords.
C. pringlei Davenp. Pringle's Lip Fern.
C. pyramidalis Fee
    var. *arizonica* (Maxon) Broun.
C. tomentosa Link. Woolly Lip Fern.
C. villosa Davenp. ex Maxon. Hairy Lip Fern.
C. wootoni Maxon. Beaded Lip Fern.
C. wrightii Hook. Wright's Lip Fern.

CYRTOMIUM.

Cyrtomium auriculatum (Underw.) Morton.

CYSTOPTERIS. Bladder Fern.

Cystopteris bulbifera (L.) Bernh. Bulblet Bladder Fern.
C. fragilis (L.) Bernh. Fragile Bladder Fern.
C. fragilis (L.) Bernh.
 var. *tenuifolia* (Clute) Broun.

DRYOPTERIS. Shield Fern, Wood Fern.

Dryopteris arguta (Kaulf.) Watt. Coastal Wood Fern.
D. filix-mas (L.) Schott. Male Fern.
D. patula (Swartz) Underw.
 var. *rossii* C. Chr.

GYMNOCARPIUM. Oak Fern.

Gymnocarpium dryopteris (L.) Newm.

NOTHOLAENA. Cloak Fern.

Notholaena aschenborniana Klotzsch.
N. aurea (Poir.) Desv.
N. californica D.C. Eaton. California Cloak Fern.
N. cochinensis Goodd. (*N. sinuata* (Sw.) Kaulf.
 var. *cochinensis* (Goodd.) Weatherby). Helechillo (16).
N. grayi Davenp.
N. integerrima Hook. (*N. sinuata* (Sw.) Kaulf.
 var. *integerrima* Hook.) (16).
N. jonesii Maxon (*Pellaea jonesii* (Maxon) Morton).
 Jones' Cloak Fern (15).
N. lemmoni D.C. Eaton.
N. limitanea Maxon (*Pellaea limitanea* (Maxon) Morton) (15).
N. neglecta Maxon.
N. parryi D.C. Eaton (*Cheilanthes parryi* (D.C. Eaton) Domin).
 Parry's Cloak Fern (15).
N. sinuata (Sw.) Kaulf. Wavy Cloak Fern.
N. standleyi Maxon.

PELLAEA. Cliff Brake.

Pellaea atropurpurea (L.) Link. Purple Cliff Brake.
P. glabella Mett. ex Kuhn
 var. *simplex* Butters (*P. atropurpurea* (L.) Link
 var. *simplex* (Butters) Morton) (17).
P. intermedia Mett. ex Kuhn.

4

P. ternifolia (Cav.) Link
   var. *wrightiana* (Hook.) A. F. Tryon (*P. wrightiana*
   Hook.) (17).
P. truncata Goodding (*P. longimucronata* Hook.) (14).

PITYROGRAMMA. Gold Fern.

Pityrogramma triangularis (Kaulf.) Maxon
   var. *maxonii* Weatherby. Goldenback Fern.

POLYPODIUM. Polypody.

Polypodium hesperium Maxon (*P. vulgare* L.
   var. *columbianum* Gilbert). Western Polypody (14).
P. thyssanolepis Klotzsch.

POLYSTICHUM. Holly Fern.

Polystichum lonchitis (L.) Roth. Holly Fern.
P. scopulinum (D.C. Eaton) Maxon. Eaton's Fern.

PTERIDIUM. Bracken, Brake.

Pteridium aquilinum (L.) Kuhn
   var. *pubescens* Underw. Western Bracken.

THELYPTERIS.

Thelypteris puberula (Baker) Morton
   var. *sonorensis* A. Reid Smith (18).

WOODSIA.

Woodsia mexicana Fee.
W. mexicana Fee
   forma pusilla Fourn.
W. oregana D.C. Eaton. Oregon Woodsia.
W. plummerae Lemmon. Flower Cup Fern.
W. scopulina D.C. Eaton. Mountain Cliff Fern.

WOODWARDIA. Chain Fern.

Woodwardia fimbriata J. E. Smith. Giant Chain Fern.

SPERMATOPHYTA.

GYMNOSPERMAE.

PINACEAE. Pine Family.

ABIES. Fir.

Abies concolor (Gordon & Glendinning) Hoopes. White Fir.
A. lasiocarpa (Hook.) Nutt. Subalpine Fir, Alpine Fir.
A. lasiocarpa (Hook.) Nutt.
    var. *arizonica* (Merriam) Lemmon. Corkbark Fir.

PICEA. Spruce.

Picea engelmannii Parry. Engelmann Spruce.
P. pungens Engelm. Blue Spruce.

PINUS. Pine.

Pinus aristata Engelm. Bristlecone Pine.
P. cembroides Zucc. Mexican Pinyon, Pino, Pinonero.
P. edulis Engelm. Pinyon, Pinyon Pine.
P. engelmannii Carr. Arizona Longleaf Pine, Apache Pine.
P. flexilis James. Limber Pine.
P. leiophylla Schiede & Deppe
    var. *chihuahana* (Engelm.) Shaw. Chihuahua Pine.
P. monophylla Torr. & Frém. Single-leaved Pinyon.
P. ponderosa Lawson
    var. *arizonica* (Engelm.) Shaw. Arizona Pine.
    var. *scopulorum* Engelm. Ponderosa Pine.
P. strobiformis Engelm. Southwestern White Pine, Pino Enano.

PSEUDOTSUGA. Douglas Fir.

Pseudotsuga menziesii (Mirb.) Franco
    var. *glauca* (Beissn.) Franco. Rocky Mountain Douglas Fir.

CUPRESSACEAE. Cypress Family.

CUPRESSUS. Cypress.

Cupressus arizonica Greene. Arizona Cypress, Cedro Blanco.
C. glabra Sudw. Smooth-barked Arizona Cypress.

JUNIPERUS. Juniper.

Juniperus californica Carr. California Juniper (19, 34).
J. *communis* L.
  var. *depressa* Pursh (*J. communis* L. var. *saxatilis* Pall.).
    Common Juniper (20).
J. *deppeana* Steud. Alligator Juniper.
J. *monosperma* (Engelm.) Sarg. One-seed Juniper.
J. *osteosperma* (Torr.) Little. Utah Juniper.
J. *scopulorum* Sarg. Rocky Mountain Juniper.

EPHEDRACEAE. Joint-Fir Family.

EPHEDRA. Joint-Fir.

Ephedra x arenicola Cutler (*E. cutleri x torreyana*).
E. *aspera* Engelm. (*E. nevadensis* Wats. var. *aspera* (Engelm.)
    Benson). Boundary Ephedra, Popotillo (21).
E. *cutleri* Peebles.
E. *fasciculata* A. Nels.
E. *nevadensis* Wats. Nevada Joint-fir.
E. *torreyana* Wats. Torrey Ephedra, Torrey Joint-fir.
E. *trifurca* Torr. Long-leaved Joint-fir, Canatilla, Popotilla,
    Teposote.
E. *viridis* Cov. Mountain Joint-fir.

ANGIOSPERMAE.

MONOCOTYLEDONEAE.

TYPHACEAE. Cat-Tail Family.

TYPHA. Cat-Tail.

Typha angustifolia L. Narrow-leaved Cat-tail (22).
T. *domingensis* Pers. Southern Cat-tail.
T. *latifolia* L. Broad-leaved Cat-tail, Espadilla.

SPARGANIACEAE. Bur-Reed Family.

SPARGANIUM. Bur-Reed.

Sparganium emersum Rehm.
  var. *multipedunculatum* (Morong) Reveal (*S. angustifolium*
    Michx.) (23, 25).
S. *eurycarpum* Engelm. Broad-fruited Bur-reed.
*S. *minimum* (Hart.) Fries (25).

7

POTAMOGETONACEAE. Pondweed Family.

POTAMOGETON.

Potamogeton crispus L. Curled Pondweed.
P. diversifolius Raf.
*P. filiformis Pers. (25).
*P. filiformis Pers.
    var. *latifolius* (Robbins) Reveal (*P. latifolius* (Robbins) Morong). Western Pondweed (25, 26).
P. foliosus Raf. (incl. vars. *foliosus* and *marcellus* Fern.). Leafy Pondweed (26).
P. gramineus L.
    var. *gramineus* Ogden.
    var. *maximus* Morong.
P. illinoensis Morong.
P. natans L. Broad-leaved Pondweed.
P. nodosus Poir.
P. pectinatus L. Sago Pondweed.
P. pusillus L.
*P. pusillus L.
    var. *tenuissimus* Mert. & Koch (*P. berchtoldii* Fieb.
    var. *tenuissimus* (Mert & Koch) Fern.) (25, 26).
*P. richardsonii (Benn.) Rydb. (22).

RUPPIACEAE.

RUPPIA.

*Ruppia maritima L. Widgeon Grass (23, 28).

ZANNICHELLIACEAE.

ZANNICHELLIA. Horned Pondweed.

Zannichellia palustris L. Common Poolmat.

NAJADACEAE. Naiad Family, Water Nymph Family.

NAJAS. Naiad, Water Nymph.

Najas flexilis Rostk. & Schmidt (27).
*N. guadalupensis Morong. Common Water Nymph (28).
N. marina L. Holly-leaved Water Nymph.

JUNCAGINACEAE. Arrow-Grass Family.

TRIGLOCHIN. Arrow-Grass.

Triglochin concinna Davy
    var. *debilis* J. T. Howell (*T. debilis* (Jones) Löve & Löve)
    (29).
T. maritima L.

ALISMATACEAE. Water-Plantain Family.

ALISMA. Water-Plantain, Mud-Plantain.

Alisma gramineum Gmel.
    var. *angustissimum* (DC.) Hendricks (14, 25).
A. triviale Pursh (incl. *A. subcordatum* Raf. sensu K. & P.) (14, 25).

ECHINODORUS. Burhead.

Echinodorus rostratus (Nutt.) Engelm. (*E. berteroi* (Sprengel)
    Fassett) (26).

SAGITTARIA. Arrow-Head.

Sagittaria cuneata Sheld.
S. graminea Michx.
S. greggii J. G. Smith.
S. latifolia Willd. Wapato, Duck Potato.
S. longiloba Engelm. Flecha de Agua.

HYDROCHARITACEAE. Frogs-Bit Family.

EGERIA.

Egeria densa Planch. (*Elodea densa* (Planch.) Caspary) (24).

ELODEA. Water-Weed.

*\ *Elodea bifoliata* St. John (25).
E. canadensis Michx.

VALLISNERIA. Tape-Grass, Eel Grass.

Vallisneria americana Michx. Water Celery.

GRAMINEAE. Grass Family.

AEGILOPS. Goat Grass.

Aegilops cylindrica Host. Jointed Goat Grass.

AEGOPOGON.

Aegopogon tenellus (DC.) Trin.
A. tenellus (DC.) Trin.
    var. *abortivus* (Fourn.) Beetle.

AGROPYRON. Wheatgrass.

Agropyron arizonicum Scribn. & Smith.
A. dasystachyum (Hook.) Scribn. Thickspike Wheatgrass.
A. desertorum (Fisch.) Schult.
A. pseudorepens Scribn. & Smith. False Quackgrass.
A. repens (L.) Beauv. Quackgrass.
A. riparian Scribn. & Smith. Streambank Wheatgrass.
A. saundersii (Vasey) Hitchc.
A. saxicola (Scribn. & Smith) Piper.
A. scribneri Vasey. Spreading Wheatgrass.
A. smithii Rydb. Western Wheatgrass.
A. smithii Rydb.
    var. *molle* (Scribn. & Smith) M. E. Jones.
    var. *palmeri* (Scribn. & Smith) Heller.
A. spicatum (Pursh) Scribn. & Smith. Bluebench Wheatgrass.
A. subsecundum (Link) Hitchc. Bearded Wheatgrass.
A. trachycaulum (Link) Malte. Slender Wheatgrass.

AGROSTIS. Bentgrass, Bent.

Agrostis exarata Trin. Spike Bent.
A. idahoensis Nash. Idaho Red Top.
A. scabra Willd.
A. semiverticillata (Forsk.) C. Chr. Water Bent.
A. stolonifera L. (A. alba L.) Red Top (26).
A. stolonifera L.
    var. *palustris* (Huds.) Farw. (*A. palustris* Huds.) Creeping
    Bent (26).

ALOPECURUS. Foxtail.

Alopecurus aequalis Sobol. Short-awn Foxtail.
A. carolinianus Walt. Carolina Foxtail.
A. geniculatus L. Water Foxtail.

ANDROPOGON. Beardgrass, Bluestem.

Andropogon gerardi Vitm. Big Bluestem.
A. gerardi Vitm.
    var. *paucipilus* (Nash) Fern. (*A. hallii* Hack.) Sand Bluestem
    (30, 31).
A. glomeratus (Walt.) B.S.P. Bushy Beardgrass.

10

ARISTIDA. Three-Awn.

Aristida adscensionis L. Six Weeks Three-awn.
A. *arizonica* Vasey. Arizona Three-awn.
A. *barbata* Fourn. Havard Three-awn.
A. *californica* Thurb.
A. *divaricata* H. & B. Poverty Three-awn.
A. *fendleriana* Steud. Fendler's Three-awn.
A. *glabrata* (Vasey) Hitchc.
A. *glauca* (Nees) Walp. Reverchon Three-awn.
A. *hamulosa* Henr.
A. *longiseta* Steud.
    var. *rariflora* Hitchc. Red Three-awn.
    var. *robusta* Merr.
A. *oligantha* Michx. Prairie Three-awn.
A. *orcuttiana* Vasey. Beggar-tick Three-awn.
A. *pansa* Woot. & Standl. Wooton Three-awn.
A. *parishii* Hitchc.
A. *purpurea* Nutt. Purple Three-awn.
A. *purpurea* Nutt.
    var. *laxiflora* Merr.
A. *ternipes* Cav. Spider Grass.
A. *ternipes* Cav.
    var. *minor* (Vasey) Hitchc.
A. *wrightii* Nash.

ARRHENATHERUM.

Arrhenatherum elatius (L.) Presl. Tall Oat Grass.

ARUNDO. Giant Reed.

Arundo donax L. Giant Reed, Carrizo.

AVENA. Oat.

Avena barbata Brot. Slender Oat.
A. *fatua* L. Wild Oat.

BECKMANNIA. Sloughgrass.

Beckmannia syzigachne (Steud.) Fern. American Sloughgrass.

BLEPHARONEURON.

Blepharoneuron tricholepis (Torr.) Nash. Hairy Drop-seed.

11

BOTHRIOCHLOA (49).

Bothriochloa barbinodis (Lag.) Herter (*Andropogon barbinodis* Lag.).

B. ischaemum (L.) Keng. (*Andropogon ischaemum* L.).

B. saccharoides (Swartz) Rydb. (*Andropogon saccharoides* Swartz). Silver Beardgrass.

B. wrightii (Hack.) Henr. (*Andropogon wrightii* Hack.).

BOUTELOUA. Grama.

Bouteloua aristidoides (H.B.K.) Grisb. Six-weeks Needle Grama, Needle Grama.

B. aristidoides (H.B.K.) Griseb.
   var. *arizonica* M. E. Jones.

B. barbata Lag. Six-weeks Grama.

B. chondrosioides (H.B.K.) Benth.

B. curtipendula (Michx.) Torr. Side Oats Grama (32).
   var. *curtipendula*
   var. *caespitosa* Gould & Kapadia.

B. eludens Griffiths.

B. eriopoda (Torr.) Torr. Black Grama.

B. glandulosa (Cerv.) Swallen.

B. gracilis (H.B.K.) Lag. ex Steud. Blue Grama.

B. hirsuta Lag. Hairy Grama.

B. parryi (Fourn.) Griffiths. Parry Grama.

B. radicosa (Fourn.) Griffiths.

B. repens (H.B.K.) Scribn. & Merr. (*B. filiformis* (Fourn.) Griffiths). Slender Grama (33).

B. rothrockii Vasey. Rothrock Grama.

B. simplex Lag. Mat Grama.

B. trifida Thurb. Red Grama.

BROMUS. Bromegrass.

Bromus arizonicus (Shear) Stebbins. Arizona Brome.

B. arvensis L. Field Brome.

B. carinatus H. & A. California Brome.

B. commutatus Schrad. Hairy Chess.

B. frondosus (Shear) Woot. & Standl.

B. inermis Leyss. Smooth Brome.

B. japonicus Thurb. Japanese Chess.

B. lanatipes (Shear) Rydb.

B. lanatipes (Shear) Rydb.
   forma *glaber* Wagnon.

B. madritensis L.

B. marginatus Nees.

12

B. mollis L. Soft Chess.
B. orcuttianus Vassey (?).
B. polyanthus Scribn. (34).
B. porteri (Coult.) Nash.
B. racemosus L.
B. richardsonii Link.
B. rigidus Roth. Ripgut Grass.
B. rubens L. Red Brome, Foxtail Chess.
B. secalinus L. Chess.
B. tectorum L. Downy Chess.
B. tectorum L.
    var. *glabratus* Spenner.
B. trinii Desv. Chilian Chess.
B. trinii Desv.
    var. *excelsus* Shear.
B. wildenowii Kunth (*B. catharticus* Vahl). Rescue Grass,
    Rescue Brome (26).

BUCHLOË. Buffalo Grass.

Buchloë dactyloides (Nutt.) Engelm.

CALAMOGROSTIS. Reed Grass.

Calamogrostis canadensis (Michx.) Beauv. Blue Joint.
C. inexpansa Gray. Northern Reed Grass.
C. scopulorum Jones.

CALAMOVILFA. Sand-Reed.

Calamovilfa gigantea (Nutt.) Scribn. & Merr. Big Sand-reed.

CATHESTECUM.

Cathestecum erectum Vasey & Hack. False Grama (?).

CENCHRUS. Sandbur (35).

Cenchrus echinatus L. Southern Sandbur.
C. insertus M. A. Curtis (*C. pauciflorus* Benth.). Field Sandbur.

CHASMANTHIUM.

Chasmanthium latifolium (Michx.) Yates (*Uniola latifolia* Michx.).
    Broadleaf Uniola, Inland Sea Oats (36).

CHLORIS. Fingergrass.

Chloris chloridea (Presl) Hitchc.
C. cucullata Bisch. Hooded Fingergrass, Hooded Windmill Grass
    (37).

CHLORIS. Fingergrass. *(cont.)*

C. gayana Kunth. Rhodes Grass.
C. latisquamea Nash.
C. verticillata Nutt. Tumble Windmill Grass, Windmill Grass.
C. virgata Swartz. Feather Fingergrass.

COTTEA. Cotta Grass.

Cottea pappaphoroides Kunth.

CYNODON. Bermuda Grass.

Cynodon dactylon (L.) Pers. Pata de Gallo.

DACTYLIS. Orchard Grass.

Dactylis glomerata L.

DACTYLOCTENIUM. Crowfoot Grass.

Dactyloctenium aegyptium (L.) Beauv.

DANTHONIA. Oat Grass.

Danthonia californica Boland. California Oatgrass.
D. intermedia Vasey. Timber Oatgrass.

DESCHAMPSIA. Hair Grass.

Deschampsia caespitosa (L.) Beauv. Tufted Hairgrass.
D. danthonioides (Trin.) Munro. Annual Hairgrass.
D. elongata (Hook.) Munro. Slender Hairgrass.

DICHANTHELIUM (38).

Dichanthelium lanuginosa (Ell.) Gould (*Panicum tennesseense* Ashe, *P. huachucae* Ashe and var. *fasciculatum* (Torr.) Hubbard).
D. oligosanthes (Schult.) Gould
var. *scribnerianum* (Nash) Gould (*Panicum scribnerianum* Nash).

DIGITARIA. Crabgrass.

Digitaria adscendens (H.B.K.) Henr. (39).
D. ischaemum (Schreb.) Schreb. Smooth Crabgrass.
D. sanguinalis (L.) Scop. Common Crabgrass.

DISTICHLIS. Saltgrass.

Distichlis spicata (L.) Greene
var. *stricta* (Torr.) Beetle (*D. stricta* (Torr.) Beetle).
Desert Saltgrass (40).

14

ECHINOCHLOA. Cock Spur.

Echinochloa colonum (L.) Link. Jungle Rice.
E. crusgalli (L.) Beauv. (incl. var. *zelayensis* (H.B.K.) Hitchc.
and var. *mitis* (Pursh) Peterm.). Barnyard Grass (26).

ELEUSINE. Goose Grass, Yard Grass.

Eleusine indica (L.) Gaertn. Goose Grass.

ELYMUS. Wild Rye.

Elymus canadensis L. Canada Wild Rye.
E. canadensis L.
    var. *robustus* (Scribn. & Smith) Mack. & Bush.
E. cinereus Scribn. & Merr.
E. glaucus Buckl. Blue Wild Rye.
E. salinus M. E. Jones. Salina Wild Rye.
E. triticoides Buckl. Beardless Wild Rye.
E. virginicus L. Virginia Wild Rye.

ELYONURUS.

Elyonurus barbiculmis Hack.

ENNEAPOGON.

Enneapogon desvauxii Beauv. Spike Pappusgrass.

ERAGROSTIS. Lovegrass.

Eragrostis arida Hitchc.
E. barrelieri Daveau.
E. chloromelas Steud. (41).
E. cilianensis (All.) Mosher. Stink Grass.
E. curvula (Schrad.) Nees. Weeping Lovegrass.
E. diffusa Buckl.
E. echinocloidea Stapf.
E. intermedia Hitchc. Plains Lovegrass.
E. lehmanniana Nees. Lehman Lovegrass.
E. lutescens Scribn.
E. mexicana (Hornem.) Link. Mexican Lovegrass.
E. neomexicana Vasey.
E. obtusiflora (Fourn.) Scribn.
E. orcuttiana Vasey.
E. pectinacea (Michx.) Nees.
E. pilosa (L.) Beauv. India Lovegrass.
E. poaeoides Beauv. (42).
E. spectabilis (Pursh) Steud. Purple Lovegrass.
E. tephrosanthos Schult. Gulf Lovegrass (41).

ERIANTHUS. Plume Grass.

Erianthus ravennae (L.) Beauv. Ravenna Grass.

ERIOCHLOA. Cup Grass.

Eriochloa aristata Vasey.
E. *contracta* Hitchc. Prairie Cupgrass.
E. *lemmoni* Vasey & Scribn. (31).
   var. *lemmoni* (incl. E. *gracilis* (Fourn.) Hitchc. var. *minor*
   (Vasey) Hitchc.). Small Southwestern Cupgrass.
   var. *gracilis* (Fourn.) Gould (E. *gracilis* (Fourn.) Hitchc.).
E. *procera* (Retz) C. E. Hubbard.

ERIONEURON (43).

Erioneuron avenaceum (H.B.K.) Tateoka var. *grandiflorum*
   (Vasey) Gould (*Tridens grandiflorus* (Vasey) Woot. &
   Standl.). Large-flowered Tridens.
E. *pilosum* (Buckl.) Nash (*Tridens pilosus* (Buckl.) Hitchc.).
   Hairy Tridens.
E. *pulchellum* (H.B.K.) Tateoka (*Tridens pulchellus* (H.B.K.)
   Hitchc.). Fluff Grass.

FESTUCA. Fescue.

Festuca arizonica Vasey. Arizona Fescue.
F. *eastwoodae* Piper.
F. *grayi* (Abrams) Piper.
F. *idahoensis* Elmer. Idaho Fescue.
F. *ovina* L. Sheep Fescue.
F. *ovina* L.
   var. *brachyphylla* (Schult.) Piper.
F. *pacifica* Piper.
F. *pratensis* Huds. (F. *elatior* L.). Meadow Fescue (26).
Festuca reflexa Buckl.
F. *rubra* L. Red Fescue.
F. *sororia* Piper.
F. *thurberi* Vasey. Thurber Fescue.

GASTRIDIUM.

Gastridium ventricosum (Gouan) Schinz & Thell.

GLYCERIA. Manna Grass.

Glyceria borealis (Nash) Batchelder. Northern Mannagrass.
G. *elata* (Nash) Hitchc. Tall Mannagrass.
G. *grandis* Wats. American Mannagrass.
G. *striata* (Lam.) Hitchc. Fowl Mannagrass.

16

HACKELOCHLOA.

Hackelochloa granularis (L.) Kuntze.

HETEROPOGON. Tangle Head.

Heteropogon contortus (L.) Beauv. Tangle Head.
H. melanocarpus (Ell.) Benth. Sweet Tangle Head.

HIEROCHLOË. Sweet Grass.

Hierochloë odorata (L.) Beauv. Sweet, Holy, Vanilla, and
Seneca Grass.

HILARIA.

Hilaria belangeri (Steud.) Nash. Curly Mesquite Grass.
H. belangeri (Steud.) Nash
   var. *longifolia* (Vasey) Hitchc.
H. jamesii (Torr.) Benth. Galleta.
H. mutica (Buckl.) Benth. Tobosa Grass, Tobosa.
H. rigida (Thurb.) Benth. Big Galleta.

HOLCUS. Velvet Grass.

Holcus lanatus L.

HORDEUM. Barley.

Hordeum arizonicum Covas.
H. brachyantherum Nevski. Meadow Barley.
H. geniculatum All. (*H. hystrix* Roth). Meditteranean Barley
   (26, 44).
H. glaucum All. (*H. stebbinsii* Covas) (26, 44).
H. jubatum L. Fox-tail Barley.
H. jubatum L.
   var. *caespitosum* (Scribn.) Hitchc.
H. leporinum Link. Wild Barley.
H. pusillum Nutt. Little Barley.
H. vulgare L. Common Barley.

IMPERATA.

Imperata brevifolia Vasey. Satintail.

KOELERIA. June Grass.

Koeleria pyramidata (Lam.) Beauv. (*K. cristata* (L.) Pers.)
   (26, 30).

LAMARCKIA. Golden Top.

Lamarckia aurea (L.) Moench.

17

LEERSIA. Cut Grass.

Leersia oryzoides (L.) Swartz. Rice Cutgrass.

LEPTOCHLOA. Sprangletop.

Leptochloa dubia (H.B.K.) Nees. Green Sprangletop.
L. fascicularis (Lam.) Gray. Beaded Sprangletop.
L. filiformis (Lam.) Beauv. Red Sprangletop.
L. uninervia (Presl) Hitchc. & Chase. Mexican Sprangletop.
L. viscida (Scribn.) Beal.

LEPTOLOMA. Fall Witch Grass.

Leptoloma cognatum (Schult.) Chase.

LOLIUM. Rye Grass.

Lolium multiflorum Lam.
L. perenne L.
L. rigidum Gaudin.
L. temulentum L. Darnel.

LYCURUS. Wolf-Tail.

Lycurus phleoides H.B.K.

MELICA. Melic Grass.

Melica frutescens Scribn.
M. nitens (Scribn.) Piper. Three-flower Melic (?).
M. porteri Scribn. Porter Melic.
M. porteri Scribn.
    var. *laxa* Boyle.

MICROCHLOA.

Microchloa kunthii Desv.

MUHLENBERGIA. Muhly.

Muhlenbergia andina (Nutt.) Hitchc. Foxtail Muhly.
M. appressa C. O. Goodding.
M. arenacea (Buckl.) Hitchc. Ear Muhly.
M. arenicola Buckl. Sand Muhly.
M. arizonica Scribn.
M. asperifolia (Nees & Mey.) Parodi. Scratchgrass.
M. brevis C. O. Goodding.
M. curtifolia Scribn.
M. depauperata Scribn.
M. dubia Fourn. Pine Muhly (?).
M. dubioides C. O. Goodding.

18

M. dumosa Scribn.

M. eludens C. G. Reeder.

M. emersleyi Vasey. Bullgrass.

M. filiculmis Vasey. Slimstem Muhly.

M. filiformis (Thurb.) Rydb. Pull-up Muhly.

M. fragilis Swallen.

M. glauca (Nees) Mez.

M. longiligula Hitchc. Long-tongue Muhly.

M. mexicana (L.) Trin.

M. mexicana (L.) Trin.
    forma setiglumis (Wats.) Fern.

M. microsperma (DC.) Kunth. Littleseed Muhly.

M. minutissima (Steud.) Swallen (*M. texana* Buckl.,
    M. sinuosa Swallen) (31).

M. montana (Nutt.) Hitchc. Mountain Muhly.

M. monticola Buckl. Mesa Muhly.

M. pauciflora Buckl. New Mexican Muhly.

M. pectinata C. O. Goodding.

M. polycaulis Scribn. Cliff Muhly.

M. porteri Scribn. Bush Muhly.

M. pulcherrima Scribn.

M. pungens Thurb. Sandhill Muhly.

M. racemosa (Michx.) B.S.P.

M. repens (Presl) Hitchc. Creeping Muhly.

M. richardsonis (Trin.) Rydb. Mat Muhly.

M. rigens (Benth.) Hitchc. Deer Grass.

M. rigida (H.B.K.) Kunth. Purple Muhly.

M. sylvatica Torr.

M. tenuifolia (H.B.K.) Trin.

M. thurberi Rydb.

M. torreyi (Kunth) Hitchc. Ring-grass.

M. utilis (Torr.) Hitchc. Aparejo Grass.

M. virescens (H.B.K.) Kunth. Screwleaf Muhly.

M. wolfii (Vasey) Rydb.

M. wrightii Vasey. Spike Muhly.

M. xerophila C. O. Goodding.

MUNROA. False Buffalo Grass.

Munroa squarrosa (Nutt.) Torr.

ORYZOPSIS. Rice Grass.

Oryzopsis bloomeri (Boland) Ricker = x *Stiporyzopsis bloomeri*
(Boland) B. L. Johnson (26).

O. hymenoides (R. & S.) Ricker. Indian Ricegrass.

O. micrantha (Trin. & Rupr.) Thurb. Littleseed Ricegrass.

O. milacea (L.) Benth. & Hook. Smilo Grass.

PANICUM.

Panicum antidotale Retz. Blue Panic.

P. arizonicum Scribn. & Merr. Arizona Panicum.

P. bulbosum H.B.K. Bulb Panicum.

P. bulbosum H.B.K.
    var. *minus* Vasey.

P. capillare L.
    var. *occidentale* Rydb. Witchgrass.

P. dichotomiflorum Michx. Fall Panicum.

P. fasciculatum Swartz
    var. *reticulatum* (Torr.) Beal. Browntop Panicum.

P. hallii Vasey. Hall's Panicum.

P. hirticaule Presl.

P. lepidulum Hitchc. & Chase.

P. miliaceum L. Broom-corn Millet.

P. obtusum H.B.K. Vine Mesquite.

P. pampinosum Hitchc. & Chase.

P. plenum Hitchc. & Chase.

P. sonorum Beal.

P. stramineum Hitchc. & Chase.

P. texanum Buckl. Texas Millet.

P. urvilleanum Kunth (?).

P. virgatum L. Switchgrass.

PAPPOPHORUM. Pappus Grass.

Pappophorum bicolor Fourn.

P. vaginatum Buckl. (*P. mucronulatum* Nees) (31).

PASPALUM.

Paspalum dilatatum Poir. Dallis Grass.

P. distichum L. Knotgrass.

P. setaceum Michx.
    var. *stramineum* (Nash) D. Banks (*P. stramineum* Nash)
    (45).

P. virletii Fourn.

PENNISETUM.

Pennisetum ciliare (L.) Link.
P. setaceum (Forsk.) Chiov. Fountain Grass.

PHALARIS. Canary Grass.

Phalaris angusta Nees.
*P. *aquatica* L. (*P. stenoptera* Hack.). Harding Grass (31, 46).
P. arundinacea L. Reed Canary Grass.
P. canariensis L. Canary Grass.
P. caroliniana Walt. Carolina Canary Grass.
P. minor Retz. Littleseed Canary Grass.
P. paradoxa L.
P. paradoxa L.
    var. *praemorsa* (Lam.) Coss. & Dur.

PHLEUM. Timothy.

Phleum alpinum L. Alpine Timothy.
P. pratense L. Common Timothy.

PHRAGMITES.

Phragmites australis (Cav.) Trin. (*P. communis* Trin.) (47).

PIPTOCHAETIUM.

Piptochaetium fimbriatum (H.B.K.) Hitchc. Pinyon Ricegrass.

POA. Blue Grass.

Poa annua L. Annual Bluegrass.
P. arida Vasey. Plains Bluegrass.
P. bigelovii Vasey & Scribn. Bigelow's Bluegrass.
P. bulbosa L. Bulbous Bluegrass.
P. canbyi (Scribn.) Piper. Canby Bluegrass.
P. compressa L. Canada Bluegrass.
P. fendleriana (Steud.) Vasey (Incl. *P. longiligula* Scribn. &
    Williams). Mutton Grass (26).
P. glaucifolia Scribn. & Williams.
P. interior Rydb. Inland Bluegrass.
*P. *juncifolia* Scribn. (25).
P. nevadensis Vasey. Nevada Bluegrass.
P. palustris L. Fowl Bluegrass.
P. pratensis L. Kentucky Bluegrass.
P. reflexa Vasey & Scribn. Nodding Bluegrass.
P. rupicola Nash. Timberline Bluegrass.
P. sandbergii Vasey (*P. secunda* Presl). Sandberg Bluegrass (26).

21

POLYPOGON.

Polypogon australis Brongn. (48).
P. *elongatus* H.B.K.
P. *interruptus* H.B.K. Ditch Polypogon.
P. *monspeliensis* (L.) Desf. Rabbitfoot Grass.

PUCCINELLIA. Alkali Grass.

Puccinellia fasciculata (Torr.) Bicknell.
P. *nuttalliana* (Schult.) Hitchc. (*P. airoides* (Nutt.) Wats. &
   Coult.) (26).
P. *parishii* Hitchc.

REDFIELDIA. Blow-Out Grass.

Redfieldia flexuosa (Thurb.) Vasey.

RHYNCHELYTRUM. Natal Grass.

Rhynchelytrum repens (Willd.) C. E. Hubb. (*R. roseus* Nees)
   (31).

SCHEDONNARDUS. Tumble Grass.

Schedonnardus paniculatus (Nutt.) Trel.

SCHISMUS.

Schismus arabicus Nees. Arabian Grass.
S. *barbatus* (L.) Thell. Meditteranean Grass.

SCHIZACHYRIUM (31).

Schizachyrium cirratum (Hack.) Woot. & Standl. (*Andropogon
   cirratus* Hack.). Texas Beardgrass.
S. *hirtiflorum* Nees (*Andropogon hirtiflorus* (Nees) Kunth).
S. *scoparium* (Michx.) Nash (*Andropogon scoparius* Michx.).
   Little Bluestem (49).
S. *scoparium* (Michx.) Nash
   var. *neomexicanum* (Nash) Gould (*Andropogon scoparius*
   Michx. var. *neomexicanus* (Nash) Hitchc.) (49).

SCLEROPOGON. Burro Grass.

Scleropogon brevifolius Phil.

SECALE. Rye.

Secale cereale L.

SETARIA. Bristle Grass.

Setaria arizonica Roem. (50).
S. faberii Herrm. (34).
S. *geniculata* (Lam.) Beauv. Knotroot Bristlegrass.
S. *glauca* (L.) Beauv. (S. *lutescens* (Weigel) Hubb.).
  Yellow Bristlegrass (51).
S. *grisebachii* Fourn. Grisebach Bristlegrass.
S. *leucopila* (Scribn. & Merr.) K. Schum.
S. *liebmanni* Fourn.
S. *macrostachya* H.B.K. Plains Bristlegrass.
S. *verticillata* (L.) Beauv. Bur Bristlegrass.
S. *villosissima* (Scribn. & Merr.) Schum.
S. *viridis* (L.) Beauv. Green Bristlegrass, Grass Foxtail.

SITANION. Squirrel Tail.

Sitanion hystrix (Nutt.) J. G. Smith.
S. *jubatum* J. G. Smith. Big Squirrel Tail.

SORGHASTRUM. Indian Grass.

Sorghastrum nutans (L.) Nash.

SORGHUM.

Sorghum bicolor (L.) Moench. (31).
S. *halapense* (L.) Pers. Johnson Grass.
S. *sudanense* (Piper) Stapf. Sudan Grass.

SPARTINA. Cord Grass.

Spartina gracilis Trin. Alkali Cordgrass.

SPHENOPHOLIS. Wedge Grass.

Sphenopholis obtusata (Michx.) Scribn. Prairie Wedgegrass.
S. *obtusata* (Michx.) Scribn.
  var. *major* (Torr.) Erdman (S. *intermedia* (Rydb.) Rydb.)
  (52).

SPOROBOLUS. Drop Seed.

Sporobolus airoides Torr. Alkali Sacaton.
S. *asper* (Michx.) Kunth. Tall Sporobolus.
S. *contractus* Hitchc. Spike Dropseed.
S. *cryptandrus* (Torr.) Gray. Sand Dropseed.
S. *flexuosus* (Thurb.) Rydb. Mesa Dropseed.
S. *giganteus* Nash. Giant Dropseed.
S. *interruptus* Vasey. Black Dropseed.
S. *nealleyi* Vasey. Nealley Dropseed.

SPOROBOLUS. Drop Seed. *(cont.)*

S. neglectus Nash.

S. patens Swallen.

S. pulvinatus Swallen.

*S. pyramidatus (Lam.) Hitchc. (50).

S. texanus Vasey.

S. vaginiflorus (Torr.) Wood.

S. wrightii Munro ex Scribn. Sacaton.

STIPA. Needle Grass.

Stipa arida Jones.

S. columbiana Macoun. Columbia Needlegrass.

S. columbiana Macoun
   var. *nelsoni* (Scribn.) Hitchc.

S. comata Trin. & Rupr. Needle and Thread.

S. comata Trin. & Rupr.
   var. *intermedia* Scribn. & Tweedy.

S. coronata Thurb.
   var. *depauperata* (Jones) Hitchc.

S. eminens Cav.

S. lettermani Vasey. Letterman Needlegrass.

S. lobata Swallen.

S. neomexicana (Thurb.) Scribn. New Mexican Feathergrass.

S. pringlei Scribn. Pringle Needlegrass.

S. robusta Scribn. Sleepygrass.

S. scribneri Vasey. Scribner Needlegrass.

S. speciosa Trin. & Rupr. Desert Needlegrass.

S. viridula Trin. Green Needlegrass.

TORREYOCHLOA.

Torreyochloa pauciflora (Presl) Church.

TRACHYPOGON. Crinkle-Awn.

Trachypogon secundus (Presl) Scribn.

TRAGUS. Bur Grass.

Tragus berteronianus Schult.

T. racemosus (L.) All. (?).

TRICHACHNE. Cotton-Top, Cotton Grass.

Trichachne californica (Benth.) Chase. Cotton-top.

T. insularis (L.) Nees.

TRICHLORIS.

Trichloris crinita (Lag.) Parodi. Feather Fingergrass.

TRIDENS.

Tridens eragrostoides (Vasey & Scribn.) Nash.
T. muticus (Torr.) Nash. Slim Tridens.
T. muticus (Torr.) Nash
    var. *elongatus* (Buckl.) Shinners (*T. elongatus* (Buckl.)
    Nash) (53).

TRIPSACUM. Gama Grass.

Tripsacum lanceolatum Rupr. Mexican Gamagrass.

TRISETUM.

Trisetum interruptum Buckl.
T. montanum Vasey.
T. spicatum (L.) Richt. Spike Trisetum.

VULPIA (54).

Vulpia myuros (L.) Gmel. (*Festuca myuros* L.).
V. myuros (L.) Gmel.
    var. *hirsuta* Hack. (*Festuca megalura* Nutt.). Foxtail Fescue.
V. octoflora (Walt.) Rydb. (*Festuca octoflora* Walt.).
    Six-weeks Fescue.

ZIZANIA. Wild Rice.

Zizania aquatica L. Annual Wildrice (55).

CYPERACEAE. Sedge Family.

BULBOSTYLIS.

Bulbostylis capillaris (L.) C. B. Clarke.
B. funckii (Steud.) C. B. Clarke.
B. juncoides (Vahl) Kükenth.
    var. *ampliceps* Kükenth.

CAREX. Sedge (26, 57).

Carex agrostoides Mack.
C. albo-nigra Mack.
C. alma Bailey.
C. aquatilis Wahl.
C. atherodes Spreng. (56).
C. athrostachya Olney.
C. aurea Nutt.
C. bella Bailey.

25

CAREX. Sedge (26, 57). *(cont.)*

C. bolanderi Olney.
C. bonplandii Kunth.
C. brevior (Dewey) Mack.
C. canescens L.
C. chalciolepis Holm.
C. chihuahensis Mack.
°C. conoidea Schk. (13).
C. curatorium Stacy.
C. disperma Dewey (57).
C. douglasii Boott.
C. ebenea Rydb.
C. eleocharis Bailey.
C. festivella Mack.
C. filifolia Nutt.
C. foenea Willd. (*C. siccata* Dewey).
C. geophila Mack.
C. hassei Bailey.
C. haydeniana Olney.
C. hystricina Muhl. Porcupine Caric-sedge, Bottle-brush
   Caric-sedge.
C. interior Bailey.
C. kelloggii W. Boott.
C. lanuginosa Michx. Woolly Sedge.
C. leptopoda Mack.
C. leucodonta Holm.
°C. mariposana Bailey (13).
C. meadii Dewey.
C. microptera Mack.
C. nebraskensis Dewey.
C. occidentalis Bailey.
C. oreocharis Holm.
C. petasata Dewey.
C. praegracilis W. Boott. Clustered Field Sedge.
C. rossii Boott.
C. rostrata Stokes. Beaked Sedge.
C. scoparia Schk.
C. senta Boott.
C. serratodens W. Boott.
C. simulata Mack. Short-beaked Sedge.
C. specuicola J. T. Howell.
C. spissa Bailey (?).
C. stipata Muhl.
C. subfusca W. Boott.

C. thurberi Dewey.
C. ultra Bailey.
C. vallicola Dewey
    var. *rusbyi* (Mack.) Herm. (*C. rusbyi* Mack.) (58).
C. vesicaria L. Inflated Sedge.
C. vulpinoidea Michx.
C. wootoni Mack.

CLADIUM. Saw Grass.

Cladium californicum (Wats.) O'Neill.

CYPERUS. Flat Sedge.

Cyperus acuminatus Torr. & Hook.
C. albomarginatus M. & S.
C. amabilis Vahl
    var. *macrostachyus* Kükenth.
C. aristatus Rottb.
C. difformis L.
C. erythrorhizos Muhl.
C. esculentus L. Chufa, Yellow Nut Sedge, Yellow Nut Grass.
C. esculentus L.
    var. *angustispicatus* Britt.
    var. *macrostachyus* Boeckl.
C. fendlerianus Boeckl.
C. hermaphroditus (Jacq.) Standl.
C. huarmensis (H.B.K.) M. C. Johnst. (*C. flavus* (Vahl) Nees)(59).
C. leavigatus L.
C. manimae H.B.K.
    var. *asperrimus* (Liebm.) Kükenth.
C. mutisii (H.B.K.) Griseb.
C. mutisii (H.B.K.) Griseb.
    var. *asper* (Liebm.) Kükenth.
C. niger R. & P.
    var. *capitatus* (Britt.) O'Neill.
C. odoratus L.
C. parishii Britt.
C. pringlei Britt.
C. rotundus L. Purple Nut Sedge, Purple Nut Grass.
C. rusbyi Britt.
C. seslerioides H.B.K.
C. spectabilis Link.
C. strigosus L.
C. uniflorus Torr. & Hook.
C. wrightii Britt.

ELEOCHARIS. Spike Rush.

Eleocharis acicularis (L.) R. & S.
E. bella (Piper) Svenson.
E. cancellata Wats. (?).
E. caribaea (Rottb.) Blake.
E. engelmanni Steud.
E. flavescens (Poir.) Urb. (25).
E. macrostachya Britt.
E. montana (H.B.K.) R. & S.
    var. *nodulosa* (Roth) Svenson.
E. montevidensis Kunth.
E. parishii Britt.
E. parvula (R. & S.) Link (25).
E. pauciflora (Lightf.) Link
    var. *suksdorfiana* (Beauverd) Svenson.
E. radicans (Poir.) Kunth.
E. rostellata (Torr.) Torr.

FIMBRISTYLIS.

Fimbristylis alamosana Fern.
F. baldwiniana (Schult.) Torr.
F. thermalis Wats.
F. vahlii (Lam.) Link.

HEMICARPHA.

Hemicarpha micrantha (Vahl) Pax.
H. micrantha (Vahl) Pax
    var. *aristulata* Coville.
    var. *drummondii* (Nees) Friedland.
    var. *minor* (Schrad.) Friedland.

SCIRPUS. Bulrush, Tule.

Scirpus acutus Muhl. Hardstem Bulrush, Great Bulrush
S. americanus Pers.
    var. *longispicatus* Britt. Three-square Bulrush.
    var. *polyphyllus* (Boeckl.) Beetle.
S. californicus (C. Meyer) Steud. Giant Bulrush.
S. microcarpus Presl.
S. olneyi Gray.
S. pallidus (Britt.) Fern.
S. paludosus A. Nels. Salt-marsh Bulrush.
S. pendulus Muhl. (42).
S. validus Vahl. Softstem Bulrush.

PALMAE. Palm Family.

WASHINGTONIA. California Palm.

Washingtonia filifera Wendl. Desert Palm, California Fan Palm.

ARACEAE. Arum Family.

PISTIA. Water-Lettuce.

Pistia stratiotes L. Water Bonnet.

LEMNACEAE. Duckweed Family.

LEMNA. Duckweed.

Lemna gibba L. Inflated Duckweed, Wind Bags.
L. minima Phil.
L. minor L. Water Lentil.
L. perpusilla Torr. (*L. aequinoctialis* Welwitsch) (60).
L. trinervis (Aust.) Small (60).
L. trisulca L. Ivy Duckweed.
L. valdiviana Phil.

SPIRODELA. Duck Mat.

Spirodela polyrhiza (L.) Schleiden.

BROMELIACEAE. Pineapple Family.

TILLANDSIA. Ball Moss.

Tillandsia recurvata L. Ball Moss, Gallitos.

COMMELINACEAE. Spiderwort.

COMMELINA. Dayflower, Widow's-Tears.

Commelina dianthifolia Delile.
C. erecta L.
    var. *crispa* (Woot.) Palmer & Steyer. Herbadel Pollo.

TRADESCANTIA. Spiderwort, Spider Lily.

Tradescantia occidentalis (Britt.) Smyth.
T. occidentalis (Britt.) Smyth
    var. *scopulorum* (Rose) Anderson & Woodson.
T. pinetorum Greene.

PONTEDERIACEAE. Pickerel Weed Family.

EICHORNIA. Water Hyacinth.

Eichornia crassipes (Mart.) Solms-Laubach (28).

HETERANTHERA. Mud Plaintain.

Heteranthera dubia (Jacq.) MacMillen. Water Star Grass.
H. limosa (Swartz) Willd.

JUNCACEAE. Rush Family.

JUNCUS. Rush.

Juncus acuminatus Michx.
J. acuminatus Michx.
    forma *sphaerocephalus* Herm.
J. acutus L.
    var. *sphaerocarpus* Engelm. Spiny Rush.
*J. articulatus L. Jointed Rush (23).
J. balticus Willd.
    var. *montanus* Engelm. Wire Rush.
*J. brevicaudatus (Engelm.) Fern. (25).
J. bufonius L. Toad Rush.
J. bufonius L.
    var. *halophilus* Buch. & Fern.
    var. *occidentalis* Herm. (*J. sphaerocarpus* Nees) (61).
J. confusus Coville.
*J. cooperi Engelm. (25).
J. drummondii E. Mey.
J. effusus L.
    var. *brunneus* Engelm. Soft Rush.
    var. *exiguus* Fern. & Wieg.
J. ensifolius Wikstr. Three-stemmed Rush.
J. interior Wieg.
J. interior Wieg.
    var. *arizonicus* (Wieg.) Herm.
    var. *neomexicanus* (Wieg.) Herm.
J. longistylis Torr.
J. longistylis Torr.
    var. *scabratus* Herm.
J. macrophyllus Coville.
J. marginatus Rostk.
J. marginatus Rostk.
    var. *setosus* Coville.
J. mertensianus Bong.

J. mexicanus Willd.
J. nevadensis Wats.
    var. *badius* (Suksd.) C. L. Hitchc. (*J. badius* Suksd.) (61).
J. saximontanus A. Nels.
J. saximontanus A. Nels.
    forma brunnescens (Rydb.) Herm.
J. tenuis Willd. Slender Rush.
J. tenuis Willd.
    var. *dudleyi* (Wieg.) Herm.
J. torreyi Coville.
J. tracyi Rydb.
J. xiphioides E. Mey.

LUZULA. Wood Rush.

Luzula multiflora (Retz) Lejeune (26).
L. parviflora (Ehrh.) Desr.
L. spicata (L.) D. C. & Lam.

LILIACEAE. Lily Family.

ALLIUM. Onion.

Allium acuminatum Hook.
A. bigelovii Wats.
A. cernuum Roth
    var. *neomexicanum* (Rydb.) Macbr. Nodding Onion.
    var. *obtusum* Ckll.
A. geyeri Wats.
A. geyeri Wats.
    var. *tenerum* Jones.
A. gooddingii Ownbey.
A. kunthii Don.
A. macropetalum Rydb.
A. nevadense Wats.
A. nevadense Wats.
    var. *cristatum* (Wats.) Ownbey.
A. palmeri Wats.
A. parishii Wats.
A. plummarae Wats.
A. rhizomatum Woot. & Standl. (incl. *A. glandulosum* Link & Otto
    sensu K. & P.).

ANDROSTEPHIUM. Funnel Lily.

Androstephium breviflorum Wats.

31

ANTHERICUM. Crag Lily, Amber Lily.

Anthericum torreyi Baker.

ASPARAGUS.

Asparagus officinalis L. Garden Asparagus.

CALOCHORTUS. Mariposa, Mariposa Lily.

Calochortus ambiguus (Jones) Ownbey.
C. flexuosus Wats. Straggling Mariposa.
C. gunnisoni Wats.
C. kennedyi Porter. Desert Mariposa.
C. kennedyi Porter
    var. *munzii* Jeps.
C. nuttallii T. & G. Sego Lily.
    var. *aureus* (Wats.) Ownbey.

DICHELOSTEMMA. Bluedicks, Coveria.

Dichelostemma pulchellum (Salisb.) Heller.
D. pulchellum (Salisb.) Heller
    var. *pauciflorum* (Torr.) Hoover.

DISPORUM.

Disporum trachycarpum (Wats.) Benth. & Hook.
D. trachycarpum (Wats.) Benth. & Hook.
    var. *subglabrum* Kelso.

EREMOCRINUM.

Eremocrinum albomarginatum Jones.

FRITILLARIA. Fritillary.

Fritillaria atropurpurea Nutt.

HESPEROCALLIS. Desert Lily.

Hesperocallis undulata Gray. Ajo.

LILIUM. Lily.

Lilium parryi Wats. Lemon Lily.
L. umbellatum Pursh (?).

MILLA. Mexican Star.

Milla biflora Cav.

NOTHOSCORDUM.

Nothoscordum texanum Jones.

32

POLYGONATUM. Solomon Seal.

Polygonatum cobrense (Woot. & Standl.) Gates.

SMILACINA.

Smilacina racemosa (L.) Desf. False Solomon Seal.

S. racemosa (L.) Desf.

    var. *amplexicaulis* (Nutt.) Wats. (26).

    var. *cylindrata* Fern. (26).

S. stellata (L.) Desf. Starflower.

STREPTOPUS. Twisted Stalk.

Streptopus amplexifolius (L.) DC.

TRITELEIA.

Triteleia lemmonae (Wats.) Greene.

TRITELEIOPSIS.

Triteleiopsis palmeri (Wats.) Hoover.

VERATRUM. False Hellebore, Corn Lily.

Veratrum californicum Durand.

ZIGADENUS. Death Camas, Poison Sego.

Zigadenus elegans Pursh. White Camas, Alkali Grass.

Z. paniculatus (Nutt.) Wats. Sand Corn.

Z. virescens (H.B.K.) Macbr.

AGAVACEAE. Agave Family.

AGAVE. Century Plant, Maguey.

*Agave arizonica Gentry & Weber (62).

A. chrysantha Peebles.

A. deserti Engelm. Desert Agave.

A. kaibabensis McKelvey.

**A. mckelveyana* Gentry (62).

A. murpheyi Gibson.

A. palmeri Engelm.

A. parryi Engelm. Parry Agave.

A. parviflora Torr.

A. schottii Engelm.

A. schottii Engelm.

    var. *treleasi* (Toumey) K. & P.

A. toumeyana Trel.

A. utahensis Engelm. Utah Agave.

33

DASYLIRION. Sotol.

Dasylirion wheeleri Wats. Desert Spoon.

NOLINA. Bear Grass.

Nolina bigelovii (Torr.) Wats. Bigelow Nolina.
N. microcarpa Wats. Sacahuista, Beargrass.
N. parryi Wats. Parry Nolina.
N. texana Wats.
      var. *compacta* (Trel.) Johnst. Bunchgrass, Sacahuista.

YUCCA. Soap Weed, Spanish Bayonet, Datil, Amole.

Yucca angustissima Engelm.
Y. arizonica McKelvey.
Y. baccata Torr. Blue Yucca, Banana Yucca, Fleshy-fruited Yucca.
Y. baccata Torr.
      var. *vespertina* McKelvey.
Y. baileyi Woot. & Standl.
Y. brevifolia Engelm. Joshua Tree.
Y. brevifolia Engelm.
      var. *jaegeriana* McKelvey.
Y. elata Engelm. Soap Tree Yucca, Palmilla.
Y. schidigera Roezl. Mohave Yucca, Spanish Dagger.
Y. schottii Engelm. Hairy Yucca.
Y. whipplei Torr. Our Lord's Candle.

AMARYLLIDACEAE. Amaryllis Family.

ZEPHYRANTHES. Zephyr Lily, Rain Lily.

Zephyranthes longifolia Hemsl. Plains Rain Lily.

HYPOXIDACEAE.

HYPOXIS. Goldeye Grass, Yellow Star Grass.

Hypoxis mexicana Schult.

IRIDACEAE. Iris Family.

IRIS.

Iris missouriensis Nutt. Rocky Mountain Iris.
I. missouriensis Nutt .
      var. *arizonica* (Dykes) R. C. Foster.
      var. *pelogonus* (Goodd.) R. C. Foster.

NEMASTYLIS.

Nemastylis tenuis (Herb.) Baker
 var. *pringlei* (Wats.) R. C. Foster.

SISYRINCHIUM. Blue-Eyed Grass.

Sisyrinchium arizonicum Rothr. Yellow-eyed Grass.
S. *cernuum* (Bickn.) Kearney.
S. *demissum* Greene. Blue-eyed Grass.
S. *demissum* Greene
 var. *amethystinum* (Bickn.) K. & P.
S. *longipes* (Bickn.) K. & P. Yellow-eyed Grass.

ORCHIDACEAE. Orchid Family.

CALYPSO.

Calypso bulbosa (L.) Oakes.

CORALLORHIZA. Coral Root.

Corallorhiza maculata Raf. Spotted Coral Root.
C. *striata* Lindl. Striped Coral Root.
C. *wisteriana* Conrad. Spring Coral Root.

CYPRIPEDIUM. Lady's Slipper.

Cypripedium calceolus L.
 var. *pubescens* (Willd.) Correll. Yellow Lady's Slipper.

EPIPACTIS. Helleborine, Stream Orchis.

Epipactis gigantea Douglas ex Hook. Giant Helleborine.

GOODYERA. Rattlesnake Plantain.

Goodyera oblongifolia Raf.
G. *repens* (L.) R. Br.

HABENARIA.

Habenaria hyperborea (L.) R. Br. Tall Northern Green Orchid.
H. *limosa* (Lindl.) Hemsl. Thurber's Bog Orchid.
H. *saccata* Greene. Slender Bog Orchid.
H. *sparsiflora* Wats. Sparsely-flowered Bog Orchid.
H. *sparsiflora* Wats.
 var. *laxiflora* (Rydb.) Correll.
*H. *viridis* (L.) R. Br.
 var. *bracteata* (Muhl.) Gray. Long-bracted Habenaria (13).

HEXALECTRIS.

Hexalectris spicata (Walt.) Barnhart. Crested Coral Root.
H. warnockii Ames & Correll. Texas Purple Spike.

LISTERA. Tway Blade.

Listera convallarioides (Swartz) Nutt. Broad-leaved Tway Blade.

MALAXIS. Adders Mouth.

Malaxis corymbosa (S. Wats.) Kuntze.
M. ehrenbergii (Reichb. f.) Kuntze.
M. soulei L. O. Williams. Mountain Malaxis.
M. tenuis (S. Wats.) Ames.

SPIRANTHES. Lady's Tresses.

Spiranthes graminea Lindl. (56).
S. michuacana (Lahlave & Lex.) Hemsl.
S. parasitica A. Rich. & Gal.
S. romanzoffiana Cham. Hooded Ladies Tresses.

DICOTYLEDONEAE.

SAURURACEAE. Lizard Tail Family.

ANEMOPSIS. Yerba-Mansa.

Anemopsis californica (Nutt.) H. & A.
    var. *subglabra* Kelso.

SALICACEAE. Willow Family.

POPULUS. Cottonwood, Poplar.

Populus acuminata Rydb. (incl. var. *rehderi* Sarg.).
    Lance-leaved Cottonwood (63).
P. angustifolia James. Narrow-leaf Cottonwood.
P. arizonica Sarg. Arizona Cottonwood, Chopo (30).
P. fremontii Wats. Fremont Cottonwood.
P. fremontii Wats.
        var. *fremontii* (incl. vars. *macdougalii* (Rose) Jeps.;
        pubescens Sarg.; *thornberi* Sarg.; *toumeyi* Sarg.) (6).
P. nigra L.
        var. *italica* Muenchh. Lombardy Poplar (?).
P. tremuloides Michx. (incl. var. *aurea* (Tidestrom) Daniels).
        Quaking Aspen, Alamo Temblon (6, 30).

SALIX. Willow (64).

Salix amygdaloides Anderss. Peach-leaf Willow.
*S. arizonica Dorn.
S. *bebbiana* Sarg. Bebb Willow.
S. *bonplandiana* H.B.K. (incl. var. *toumeyi* (Britt.) Schneid. and
    S. *laevigata* Bebb and its var. *araquipa* (Jeps.) Ball).
    Bonpland Willow.
S. *exigua* Nutt. (incl. vars. *nevadensis* (Wats.) Schneid. and
    stenophylla (Rydb.) Schneid.). Coyote Willow.
S. *geyeriana* Anderss. Geyer Willow.
S. *gooddingii* Ball (incl. var. *variabilis* Ball). Goodding Willow.
S. *irrorata* Anderss.
S. *lasiandra* Benth. Pacific Willow.
S. *lasiandra* Benth.
    var. *caudata* (Nutt.) Sudw. (S. *caudata* (Nutt.) Heller
    incl. var. *bryantiana* Ball & Bracelin).
S. *lasiolepis* Benth. (incl. var. *bracelinae* Ball). Arroyo Willow.
S. *ligulifolia* Ball (S. *lutea* Nutt. var. *ligulifolia* Ball).
    Strapleaf Willow.
S. *lutea* Nutt. ex Schneider. Yellow Willow (?).
S. *monticola* Bebb (S. *padophylla* Rydb.). Serviceberry Willow.
S. *scouleriana* Barratt ex Hook. Scouler Willow.
S. *taxifolia* H.B.K. (incl. var. *microphylla* (S. & C.) Schneid.).
    Yew Leaf Willow.
Note: Infraspecific names are included with the species pending
    further study.

JUGLANDACEAE. Walnut Family.

JUGLANS. Walnut.

Juglans major (Torr.) Heller. Arizona Walnut, Nogal Silvestre.

BETULACEAE. Birch Family.

ALNUS. Alder.

Alnus oblongifolia Torr. Arizona Alder.
A. tenuifolia Nutt. Thin-leaf Alder.

BETULA. Birch.

Betula occidentalis Hook. Water Birch.

OSTRYA. Hop Hornbeam.

Ostrya knowltoni Coville. Knowlton Hop Hornbeam.

FAGACEAE. Beech Family.

QUERCUS. Oak.

Quercus ajoensis C. H. Muell.
Q. ajoensis x turbinella.
Q. arizonica Sarg. Arizona White Oak.
Q. arizonica x *Q. gambelii* (65).
Q. arizonica x *Q. grisea.*
Q. arizonica x *Q. rugosa.*
Q. chrysolepis Liebm. Canyon Live Oak.
Q. dunnii Kell. Palmer Oak.
Q. emoryi Torr. Emory Oak, Bellota.
Q. gambelii Nutt. Gambel Oak.
Q. grisea Liebm. (*Q. oblongifolia* Torr.). Gray Oak (66).
Q. hypoleucoides Camus. Silver Leaf Oak.
Q. x pauciloba Rydb. (*Q. gambelii* x *turbinella*) (67).
Q. pungens Liebm. Sandpaper Oak.
Q. rugosa Nee. Netleaf Oak.
Q. toumeyi Sarg. Toumey Oak.
Q. turbinella Greene. Shrub Live Oak, Turbinella Oak.
Q. undulata Torr. Wavy Leaf Oak.

ULMACEAE. Elm Family.

CELTIS. Hackberry.

Celtis pallida Torr. Desert Hackberry, Granjeno.
C. reticulata Torr. Net Leaf Hackberry, Palo Blanco.

ULMUS. Elm.

Ulmus pumila L. Siberian Elm (68).

MORACEAE. Mulberry Family.

HUMULUS. Hop.

Humulus americanus Nutt.

MORUS. Mulberry.

Morus microphylla Buckl. Texas Mulberry.

URTICACEAE. Nettle Family.

PARIETARIA. Pellitory.

Parietaria hespera Hinton (*P. floridana* Nutt.). Pellitory (69).
P. pensylvanica Muhl. Hammerwort.

URTICA. Nettle.

Urtica gracilenta Greene.

U. gracilis Ait. Tall White Nettle.

U. serra Blume.

*U. urens L. Dwarf Nettle, Ortiga (28).

VISCACEAE. Mistletoe Family.

ARCEUTHOBIUM. Small Mistletoe (70, 71).

Arceuthobium abietinum Engelm.
  forma concoloris Hawksworth & Wiens (*A. campylopodum*
  Engelm. *forma abietinum* (Engelm.) Gill). White Fir Dwarf
  Mistletoe.
A. apachecum Hawksworth & Wiens. Apache Dwarf Mistletoe.
A. blumeri A. Nels. (*A. campylopodum* Engelm.
  forma blumeri (Engelm.) Gill).
A. cyanocarpum Coult. & Nels. (*A. compylopodum* Engelm.
  forma cyanocarpum (A. Nels.) Gill). Limber Pine Dwarf
  Mistletoe.
A. divaricatum Engelm. (*A. campylopodum* Engelm. *forma*
  divaricatum (Engelm.) Gill). Pinyon Dwarf Mistletoe.
A. douglasi Engelm. Douglas Fir Dwarf Mistletoe.
*A. gillii Hawksworth & Wiens (72).
A. microcarpum (Engelm.) Hawksworth & Wiens
  (*A. campylopodum* Engelm. *forma microcarpum* (Engelm.)
  Gill). Western Spruce Dwarf Mistletoe.
A. vaginatum (Willd.) Presl
  ssp. *cryptopodum* (Engelm.) Hawksworth & Wiens
  (*A. vaginatum* (H.B.K.) Eichler). Southwestern Dwarf
  Mistletoe.

PHORADENDRON. Mistletoe (73).

Phoradendron bolleanum (Seem.) Eichler
  ssp. *densum* (Torr.) Wiens.
  ssp. *pauciflorum* (Torr.) Wiens.
P. californicum Nutt. (incl. var. *distans* Trel.). Desert Mistletoe.
P. capitellatum Torr. ex Trel. (*P. bolleanum* (Seem.) Eichler
  var. *capitellatum* (Torr. ex Trel.) K. & P.).
P. juniperinum Engelm.
P. tomentosum (DC.) Gray (*P. flavescens* (Pursh) Nutt.). Inierto.
P. tomentosum (DC.) Gray
  ssp. *macrophyllum* (Engelm.) Wiens (*P. flavescens* (Pursh)
  Nutt. var. *macrophyllum* Engelm.).

39

PHORADENDRON. Mistletoe (73). *(cont.)*
P. villosum (Nutt.) Nutt.
    ssp. *coryae* (Trel.) Wiens (*P. coryae* Trel.).
    ssp. *villosum.*

SANTALACEAE. Sandlewood Family.

COMANDRA. Bastard Toadflax.
Comandra pallida A. DC.

ARISTOLOCHIACEAE. Birthwort Family.

ARISTOLOCHIA.
Aristolochia watsoni Woot. & Standl. Indian Root.

RAFFLESIACEAE. Rafflesia Family.

PILOSTYLES.
Pilostyles thurberi Gray.

POLYGONACEAE. Buckwheat Family.

CHORIZANTHE.
Chorizanthe brevicornu Torr. Brittle Spine Flower.
C. corrugata (Torr.) T. & G. Corrugated Spiny Herb.
C. rigida (Torr.) T. & G. Rigid Spiny Herb.
C. thurberi (Gray) Wats. Thurber Spiny Herb.
C. watsoni T. & G. Watson Spiny Herb.

ERIOGONUM. Wild Buckwheat (74).
Eriogonum abertianum Torr.
E. abertianum Torr.
    var. *cyclosepalum* (Greene) Fosburg.
E. alatum Torr. (*E. alatum* Torr. ssp. *triste* Stokes).
    Winged Eriogonum.
E. alatum Torr.
    var. *mogellense* Stokes.
E. apachense Reveal (75).
E. arizonicum Stokes.
E. brachypodum T. & G. (*E. parryi* Gray). Tecopa,
    Skeleton Weed.
E. caespitosum Nutt. Tufted Buckwheat.
E. capillare Small.
E. cernuum Nutt. Nodding Eriogonum.
E. corymbosum Benth.

E. corymbosum Benth.
  var. *glutinosum* (Jones) Jones (*E. aureum* Jones).
  var. *orbiculatum* (Stokes) Reveal & Brotherson.
  var. *velutinum* Reveal.
E. darrovii Kearney.
E. davidsonii Greene (*E. vimineum* Dougl. ssp. *juncinellum*
  (Gandoger) Stokes). Brown Eriogonum.
E. deflexum Torr. Skeleton Weed.
E. deflexum Torr.
  var. *turbinatum* (Small) Reveal.
E. divaricatum Hook.
E. ericifolium T. & G. (*E. mearnsii* Parry).
E. ericifolium T. & G.
  var. *pulchrum* (Eastw.) Reveal (*E. mearnsii* Parry var.
  pulchrum (Eastw.) Reveal).
E. fasciculatum Benth.
  var. *polifolium* (Benth.) T. & G.
E. gordonii Benth.
E. heermannii Dur. & Hilg.
  var. *argense* (Jones) Munz. Hermann Buckwheat.
  var. *sulcatum* (Wats.) Munz & Reveal (*E. sulcatum* Wats.)
E. hieracifolium Benth.
E. hookeri Wats.
E. inflatum Torr. & Frém. (*E. clutei* Rydb.). Desert Trumpet.
E. inflatum Torr. & Frém.
  var. *deflatum* I. M. Johnst.
E. insigne Wats. (*E. deflexum* Torr. ssp. *insigne* (Wats.) Stokes).
E. jamesii Benth. Antelope Sage.
E.jamesii Benth.
  var. *flavescens* Wats. (*E. bakeri* Greene).
  var. *undulatum* (Benth.) Stokes.
E. jonesii Wats.
E. kearneyi Wats.
E. lachnogynum Torr.
E. leptocladon T. & G.
  *   var. *papiliunculi* Reveal (76).
  var. *ramosissimum* (Eastw.) Reveal.
E. leptophyllum (Torr.) Woot. & Standl.
E. maculatum Heller. Angle-stemmed Buckwheat.
E. microthecum Nutt.
  var. *foliosum* (T. & G.) Reveal (*E. simpsonii* Benth.).
  var. *laxiflorum* Hook.
E. mortonianum Reveal (76).
E. nidularium Cov. Whisk Broom.

41

ERIOGONUM. Wild Buckwheat (74). *(cont.)*

E. ordii Wats.

E. ovalifolium Nutt. Oval-leaved Buckwheat.

E. ovalifolium Nutt.
>   var. *multiscapum* Gandoger (?).

E. palmerianum Reveal.

E. parishii Wats.

E. pharnaceoides Torr.

E. pharnaceoides Torr.
>   var. *cervinum* Reveal.

E. plumatella Dur. & Hilg. Flat Top.

E. polycladon Benth. Sorrel Erigonum.

E. pusillum T. & G. Yellow Turban.

E. racemosum Nutt. Red Root Buckwheat.

E. reniforme Torr. & Frém. Kidney-leaved Buckwheat.

E. ripleyi J. T. Howell.

E. rotundifolium Benth.

E. shockleyi Wats.
>   var. *longilobum* (Jones) Reveal.

E. subreniform Wats.

E. thomasii Torr. Thomas Eriogonum.

E. thompsonae Wats.

E. thompsonae Wats.
>   var. *atwoodii* Reveal.

E. thurberi Torr.

E. trichopes Torr. Little Trumpet.

E. umbellatum Torr.
>   var. *umbellatum*
>   var. *cognatum* (Greene) Reveal (*E. cognatum* Greene).
>   var. *subaridum* Stokes.

E. wetherillii Eastw.

E. wrightii Torr. Wright Buckwheat.

E. wrightii Torr.
>   var. *pringlei* (Coult. & Fisher) Reveal (*E. pringlei* Coult.
>   & Fisher).

E. zionis J. T. Howell (?).

E. zionis J. T. Howell
>   var. *coccineum* J. T. Howell.

FAGOPYRUM. Buckwheat.

Fagopyrum sagittatum Gilib.

NEMACAULIS.

Nemacaulis denudata Nutt. Woolly Heads.

42

OXYRIA. Mountain Sorrel.

Oxyria digyna (L.) Hill.

OXYTHECA.

Oxytheca perfoliata T. & G. Punctured Bract.

POLYGONUM.

Polygonum amphibium L.
  var. *stipulaceum* Coleman. Water Smartweed.
P. *argyrocoleon* Steud. Silversheath Knotweed.
P. *aviculare* L. Prostrate Knotweed.
P. *aviculare* L.
  var. *littorale* (Link) W.D.J. Koch.
*P. *bicorne* Raf. (25).
P. *bistortoides* Pursh. Bistort.
P. *bistortoides* Pursh
  var. *oblongifolium* (Meisn.) St. John.
P. *coccineum* Muhl.
*P. *confertiflorum* Nutt. (25).
P. *convolvulus* L. Black Bindweed.
P. *douglasii* Greene.
P. *fusiforme* Greene.
*P. *hydropiperoides* Michx. (25).
P. *incanum* F.W. Schmidt.
P. *kelloggii* Greene.
P. *lapathifolium* L. Willow Smartweed.
*P. *minimum* Wats. (34).
P. *pensylvanicum* L. Pinkweed.
P. *persicaria* L. Lady's Thumb, Moco de Guajolote.
P. *punctatum* Ell. Water Smartweed.
P. *ramosissimum* Michx. Bush Knotweed.
P. *sawatchense* Small.
*P. *triandrous* Coolidge (77).

PTEROSTEGIA.

Pterostegia drymarioides Fisch. & Mey.

RUMEX. Dock, Sorrel.

Rumex acetosella L. Sheep Sorrel.
R. *altissimus* Wood. Peachleaf Dock, Palė Dock.
R. *californicus* Rech. f.
R. *conglomeratus* Murr. Cluster Dock.
R. *crispus* L. Curley Dock.

43

RUMEX. Dock, Sorrel. *(cont.)*

R. *densiflorus* Osterh.

*R. *dentatus* L. (13).

R. *ellipticus* Greene.

R. *fueginus* Phil.

R. *hymenosepalus* Torr. Canigre, Wild Rhubarb.

R. *mexicanus* Meisn. (*R. triangulivalvis* (Danser) Rech. f.) (78).

R. *nematopodus* Rech. f.

R. *obtusifolius* L. Bitter Dock.

R. *orthoneurus* Rech. f.

R. *violascens* Rech. f.

STENOGONUM (79).

Stenogonum flexum Jones (*Eriogonum flexum* Jones).

S. salsuginosum Nutt. (*Eriogonum salsuginosum* (Nutt.) Hook.).

CHENOPODIACEAE. Goose Foot Family.

ALLENROLFEA. Iodine Bush.

Allenrolfea occidentalis (Wats.) Kuntze. Pickleweed.

ATRIPLEX. Salt Bush, Orache.

Atriplex acanthocarpa (Torr.) Wats.

A. *argentea* Nutt. ssp. *typica* Hall & Clements
    var. *caput-medusae* (Eastw.) Fosberg. Silver Saltbush.

A. *canescens* (Pursh) Nutt. Wingscale, Cenizo, Chamiso.

A. *canescens* (Pursh) Nutt.
    ssp. *linearis* Hall & Clements (*A. linearis* Wats.).
    Narrow-leaved Wingscale (80).

A. *confertifolia* (Torr. & Frém.) Wats. Shad Scale.

A. *elegans* (Moq.) D. Dietr. (incl. var. *thornberi* Jones).
    Wheelscale Saltbush (80).

A. *elegans* (Moq.) D. Dietr.
    ssp. *fasciculata* (Wats.) Hall & Clements (*A. fasciculata*
    Wats.) (80).

A. *garrettii* Rydb.

A. *griffithsii* Standl.

A. *hymenelytra* (Torr.) Wats. Desert Holly.

A. *jonesii* Standl.

A. *lentiformis* (Torr.) Wats. Quail Brush, Lens Scale.

A. *obovata* Moq.

A. *patula* L.
    var. *hastata* (L.) Gray.

A. *polycarpa* (Torr.) Wats. All Scale, Cattle Spinach.

44

A. powellii Wats.

A. rosea L. Red Scale, Red Orache.

A. saccaria Wats.

A. semibacata R. Br. Australian Saltbush.

A. wrightii Wats. Wright Saltbush.

BASSIA. Smother Weed.

Bassia hyssopifolia (Pall.) Kuntze. Five Hook Bassia.

CHENOPODIUM. Goose Foot, Pigweed.

Chenopodium album L. Common Lambs Quarters.

C. ambrosioides L. ssp. *euambrosioides* Aellen

    var. *typicum* (Speg.) Aellen. Mexican Tea, Epizote, Quelite.

    var. *anthelminticum* (L.) Aellen.

C. atrovirens Rydb.

C. berlandieri Moq.

    var. *sinuatum* (Murr) Wahl. Pitseed Goosefoot.

    var. *zschackei* (Murr) Murr.

C. botrys L. Jerusalem Oak.

C. capitatum (L.) Asch. Strawberry Blite.

C. chenopoidioides (L.) Aellen (*C. rubrum* L.) (34).

C. desiccatum A. Nels. Narrowleaf Goosefoot.

C. desiccatum A. Nels.

    var. *leptophylloides* (Murr) Wahl.

C. fremontii Wats.

C. fremontii Wats.

    var. *incanum* Wats.

C. glaucum L.

    ssp. *salinum* (Standl.) Aellen.

C. graveolens Willd.

    var. *neomexicanum* (Aellen) Aellen.

C. hians Standl.

C. incognitum Wahl.

C. leptophyllum (Nutt. ex Moq.) S.Wats. Slimleaf Goosefoot (81).

C. murale L. Nettleleaf Goosefoot.

C. neomexicanum Standl.

C. palmeri Standl.

C. pumilio R. Br. Ridged Goosefoot (82).

C. watsoni A. Nels.

CORISPERMUM. Bug Seed, Tickseed.

Corispermum hyssopifolium L.

C. nitidum Kit.

45

CYCLOLOMA. Winged Pigweed.

Cycloloma atriplicifolium (Spreng.) Coult.

EUROTIA. Winter Fat.

Eurotia lanata (Pursh) Moq. Winter Fat.
E. lanata (Pursh) Moq.
    var. *subspinosa* (Rydb.) K. & P.

GRAYIA. Hop Sage.

Grayia brandegei Gray.
G. spinosa (Hook.) Moq. Spiny Hop Sage.

KOCHIA. Summer Cypress.

Kochia americana Wats. Red Molly.
K. americana Wats.
    var. *vestita* Wats.
K. scoparia (L.) Schrad. Summer Cypress Belvedere.
K. scoparia (L.) Schrad.
    var. *subvillosa* Moq.

MONOLEPIS. Patata.

Monolepis nuttalliana (Schult.) Greene. Poverty Weed.

NITROPHILA.

Nitrophila occidentalis (Moq.) Wats. Alkali Weed.

SALSOLA. Russian Thistle (83).

Salsola iberica Sennen & Pau (*S. kali* L. var. *tenuifolia* (Tausch.)
    Aellen).
S. paulsenii Litv.

SARCOBATUS. Grease Wood.

Sarcobatus vermiculatus (Hook.) Torr. Chico.

SUAEDA. Seep Weed, Sea Blight.

Suaeda calceoliformis (Hook.) Moq. (*S. depressa* (Pursh) Wats.) (84).
S. suffrutescens Wats.
S. torreyana Wats. Quelite Salado, Desert Seepweed.
S. torreyana Wats.
    var. *ramosissima* (Standl.) Munz.

ZUCKIA.

Zuckia arizonica Standl.

AMARANTHACEAE. Amaranth Family.

ACANTHOCHITON. Green Stripe.

Acanthochiton wrightii Torr.

ALTERNANTHERA. Chaff Flower.

Alternanthera repens (L.) Kuntze. Khakiweed.

AMARANTHUS. Amaranth, Pig Weed.

Amaranthus albus L. Tumble Pigweed.
A. albus L.
    var. *pubescens* (Uline & Bray) Fern.
*A. *crassipes* Schlecht (42).
A. cruentus L. Amarantobojo, Purple Amaranth.
A. fimbriatus (Torr.) Benth. Fringed Amaranth.
A. fimbriatus (Torr.) Benth.
    var. *denticulatus* (Torr.) Uline & Bray.
A. graecizans L. Prostrate Pigweed, Cochino, Quelite Manchado.
A. hybridus L. Spleen Amaranth, Quelite Morado.
A. leucocarpus Wats.
A. obcordatus (Gray) Standl.
A. palmeri Wats. Palmer's Amaranth, Bledo, Quelite.
A. powellii Wats.
A. retroflexus L. Rough Pigweed.
A. retroflexus L.
    var. *salicifolius* Johnst. Careless Weed.
A. torreyi (Gray) Benth. Torrey's Amaranth (?).
A. viridis L.
A. watsoni Standl.
A. wrightii Wats.

BRAYULINEA.

Brayulinea densa (H. & B.) Small. Small Matweed.

FROELICHIA. Snake Cotton, Cotton Weed.

Froelichia arizonica Thornber.
F. gracilis (Hook.) Moq. Slender Snake Cotton.
F. interrupta (L.) Moq. (50).

GOMPHRENA. Globe Amaranth.

Gomphrena caespitosa Torr.
G. nitida Rothr.
G. sonorae Torr.
*G. *viridis* Woot. & Standl. (50).

47

IRESINE. Bloodleaf.

Iresine heterophylla Standl.

TIDESTROMIA.

Tidestromia lanuginosa (Nutt.) Standl. Woolly Tidestromia, Espanta Vaqueras.
T. oblongifolia (Wats.) Lindl.

NYCTAGINACEAE. Four O'Clock Family.

ABRONIA. Sand Verbena.

Abronia angustifolia Greene
   var. *arizonica* (Standl.) K. & P.
A. elliptica A. Nels. (*A. pumila* Rydb.) (85).
A. fragans Nutt. Snowball, Sweet Sand Verbena.
A. nana Wats.
A. villosa Wats. Hairy Sand Verbena.

ACLEISANTHES.

Acleisanthes longiflora Gray. Angel Trumpet, Yerba-de-la-Rabia.

ALLIONIA.

Allionia choisyi Standl. Smooth Umbrella Wort.
A. cristata (Standl.) Standl.
A. incarnata L. Trailing Four-O'Clock, Windmills.

AMMOCODON. Moonpod.

Ammocodon chenopodioides (Gray) Standl. Goosefoot Moonpod.

ANULOCAULIS. Ringstem.

Anulocaulis leisolenus (Torr.) Standl.

BOERHAAVIA. Spiderling.

Boerhaavia coccinea Mill. Red Spiderling.
B. coulteri (Hook. f.) Wats. Coulter Spiderling.
B. erecta L.
B. gracillima Heimerl.
B. intermedia Jones. Five-winged Ringstem.
B. megaptera Standl.
B. pterocarpa Wats.
B. purpurascens Gray. Purple Spiderling.
B. spicata Choisy.
B. torreyana (Wats.) Standl.
B. triquetra Wats. (50).
B. wrightii Gray. Large-bracted Boerhaavia.

48

COMMICARPUS.

Commicarpus scandens L.

MIRABILIS.

Mirabilis bigelovii Gray.
M. bigelovii Gray
    var. *retrorsa* (Heller) Munz. Wishbone Bush.
*M. *froebelii* (Behr) Greene (50).
M. jalapa L. The Cultivated Four-O'Clock.
M. longiflora L. Sweet Four O'Clock.
M. longiflora L.
    var. *wrightiana* (Gray) K. & P.
M. multiflora (Torr.) Gray. Colorado Four O'Clock.
M. oxybaphoides Gray. Spreading Four O'Clock.

OXYBAPHUS.

Oxybaphus coccineus Torr.
O. comatus (Small) Weath.
O. glaber Wats.
O. linearis (Pursh) Robins.
O. linearis (Pursh) Robins.
    var. *decipiens* (Standl.) K. & P.
O. pumilis (Standl.) Standl.

SELINOCARPUS. Moonpod.

Selinocarpus nevadensis (Standl.) Fowler & Turner (86).

TRIPTEROCALYX.

Tripterocalyx carnea (Greene) Galloway
    var. *wootoni* (Standl.) Galloway (*T. wootoni* Standl.) (85).
T. micranthus (Torr.) Hook

PHYTOLACCACEAE. Pokeberry Family.

PHYTOLACCA. Pokeberry.

Phytolacca americana L. Pokeweed, Scoke.

RIVINA. Rouge Plant, Pigeon Berry.

Rivina humilis L. Coralito.

AIZOACEAE. Carpet Weed Family.

CYPSELEA.

Cypselea humifusa Turpin (55).

49

GLINUS.

Glinus radiatus (R. & P.) Rohrb. (87).

MOLLUGO. Carpet Weed.

Mollugo cerviana (L.) Seringe. Thread-stem Carpet Weed.
M. verticillata L. Indian Chick Weed.

SESUVIUM. Sea Purslane.

Sesuvium verrucosum Raf.

TRIANTHEMA. Horse Purslane.

Trianthema portulacastrum L. Verdolaga Blanca.

PORTULACACEAE. Portulaca Family.

CALANDRINIA. Rock Purslane.

Calandrinia ambigua (Wats.) Howell. Desert Pot Herb.
C. ciliata (R. & P.) DC.
   var. *menziesii* (Hook.) Macbr. Red Maids.

CALYPTRIDIUM.

Calyptridium monandrum Nutt. Sand Cress.
C. parryi Gray
   var. *arizonicum* J. T. Howell.

CLAYTONIA. Spring Beauty.

Claytonia perfoliata Donn (*Montia perfoliata* (Donn) Howell).
   Miner's Lettuce (88).
C. rosea Rydb.

LEWISIA.

Lewisia brachycalyx Engelm.
L. pygmaea (Gray) Robins.
L. rediviva Pursh. Bitter Root.

MONTIA.

Montia chamissoi (Ledeb.) Durand & Jackson.

PORTULACA. Purslane.

Portulaca mundula Johnst. Chisme.
P. oleracea L. Common Purslane, Verdolaga.
P. parvula Gray.
P. retusa Engelm. Western Pusley.
P. suffrutescens Englem.
P. umbraticola H.B.K. (*P. coronata* Small).

TALINUM.

Talinum angustissimum (Gray) Woot. & Standl.
T. aurantiacum Engelm. Flame Flower.
T. brevifolium Torr.
T. gooddingii P. Wilson.
T. paniculatum (Jacq.) Gaertn. Pink Baby Breath, Ramadel Sapo.
T. parviflorum Nutt. Dwarf Flame Flower.
T. validulum Greene.

CARYOPHYLLACEAE. Pink Family.

ACHYRONYCHIA.

Achyronychia cooperi Gray. Frost Mat.

ARENARIA. Sandwort.

Arenaria abberrans Jones.
A. confusa Rydb.
A. douglasi Fenzl.
A. eastwoodiae Rydb.
A. eastwoodiae Rydb.
    var. *adenophora* K. & P.
A. fendleri Gray.
A. fendleri Gray
    var. *brevifolia* (Maguire) Maguire.
    var. *porteri* Rydb.
    var. *tweedyi* (Rydb.) Maguire.
A. filiorum Maguire.
A. lanuginosa (Michx.) Rohrb.
    ssp. *saxosa* (Gray) Maguire.
A. macradenia Wats.
A. macradenia Wats.
    ssp. *ferrisiae* Abrams.
    var. *parishiorum* Robins. Desert Sandworm.
A. obtusiloba (Rydb.) Fern.
A. rubella (Wahl.) J. E. Smith.

CERASTIUM. Mouse-Ear Chickweed.

Cerastium arvense L.
C. beeringianum C. & S.
C. brachypodum (Engelm.) Robins.
C. nutans Raf. Powder Horn.

51

CERASTIUM. Mouse-Ear Chickweed. *(cont.)*

C. nutans Raf.
    var. *obtectum* K. & P.
C. sordidum Robins.
C. texanum Britt.
C. vulgatum L. Common Mouse-ear.

DRYMARIA. Drymary.

Drymaria depressa Greene.
D. effusa Gray.
D. fendleri Wats.
D. molluginea (Lag.) Didr. (*D. sperguloides* Gray) (89).
D. pachyphylla Woot. & Standl.
D. tenella Gray.

HERNIARIA. Burst-Wort.

Herniaria cinerea DC.

LOEFLINGIA.

Loeflingia squarrosa Nutt. (including *L. texana* Hook.) (30).

LYCHNIS. Campion.

Lychnis drummondii (Hook.) Wats.

PARONYCHIA. Nailwort.

Paronychia jamesii T. & G.
P. sessiliflora Nutt. (90).

SAGINA. Pearlwort.

Sagina occidentalis Wats.
S. saginoides (L.) Karst.
    var. *hesperia* Fern.

SAPONARIA. Soapwort.

Saponaria officinalis L. Bouncing Bet.
S. vaccaria L. Cow Soapwort.

SILENE. Catchfly, Campion.

Silene acaulis L.
    ssp. *subcaulescens* (F. N. Williams) Hitchc. & Maguire.
    Moss Campion.
S. antirrhina L. Sleepy Catchfly.
S. gallica L. Forked Catchfly.

S. laciniata Cav.
　　ssp. *greggii* (Gray) Hitchc. & Maguire. Mexican Campion.
S. menziesii Hook.
S. rectiramea Robins.
S. scouleri Hook.
　　ssp. *pringlei* (Wats.) Hitchc. & Maguire.
　　var. *typica.*
　　var. *concolor* Hitchc. & Maguire.
　　var. *eglandulosa* Hitchc. & Maguire.
　　var. *leptophylla* Hitchc. & Maguire.
S. thurberi Wats.

SPERGULARIA. Sand Spurry.

Spergularia marina (L.) Griseb. Salt Marsh Sand Spurry.

STELLARIA. Starwort, Chickweed.

Stellaria gonomischa Boivin.
S. jamesiana Torr.
S. longifolia Muhl.
S. longipes Goldie.
S. media (L.) Cyrillo. Common Chickweed.
S. nitens Nutt.

NYMPHACEAE. Waterlily Family.

NUPHAR. Spatterdock.

*Nuphar luteum (L.) Sibth. & Sm.
　　ssp. *polycepalum* (Engelm.) E. O. Beal. Yellow Pond Lily
　　(91, 92).

NYMPHAEA. Water-Lily (25).

*Nymphaea mexicana Zucc. Yellow Water-lily, Lampazo Amarillo.
*N. odorata Ait. White Water-lily, Ninfa Acuatica.

CERATOPHYLLACEAE. Hornwort, Coon Tail.

CERATOPHYLLUM.

Ceratophyllum demersum L. Common Hornwort.

RANUNCULACEAE. Crowfoot Family.

ACONITUM. Monks Hood.

Aconitum columbianum Nutt.
A. infectum Greene.

53

ACTAEA. Baneberry.

Actaea rubra (Ait.) Willd. (34, 44).
   ssp. *arguta* (Nutt.) Hult.

ANEMONE.

Anemone cylindrica Gray.
A. globosa Nutt.
A. tuberosa Rydb. Desert Windflower.

AQUILEGIA. Columbine.

Aquilegia caerulea James
   ssp. *pinetorum* (Tidestrom) Payson. Rocky Mountain
   Columbine.
A. chrysantha Gray.
A. desertorum (Jones) Ckll.
A. elegantula Greene.
A. longissima Gray. Long Spur Columbine.
A. micrantha Eastw.
A. triternata Payson.

CALTHA. Marsh Marigold.

Caltha leptosepala DC. Elks Lip.

CIMICIFUGA. Bug Bane.

Cimicifuga arizonica Wats.

CLEMATIS. Clematis, Virgin's Bower.

Clematis bigelovii Torr.
C. drummondii T. & G. Texas Virgin Bower, Barbas de Chivato.
C. hirsutissima Pursh. Leather Flower.
C. hirsutissima Pursh
   var. *arizonica* (Heller) Erickson.
C. ligusticifolia Nutt.
C. palmeri Rose.
C. pseudoalpina (Kuntze) A. Nels.

DELPHINIUM. Larkspur.

Delphinium ajacis L. Rock Larkspur, Espuela de Caballero (34).
D. andesicola Ewan.
D. andesicola Ewan
   ssp. *amplum* Ewan.
D. barbeyi Huth.
D. geraniifolium Rydb.

D. nelsoni Greene
  forma *pinetorum* (Tidestrom) Ewan.
D. parishii Gray (*D. amabile* Tidestrom incl. ssp. *apachense*
  (Eastw.) Ewan) (93).
D. scaposum Greene. Barestem Larkspur.
D. scopulorum Gray.
D. tenuisectum Greene
  ssp. *amplibracteatum* (Woot.) Ewan.
D. virescens Nutt.
  ssp. *wootoni* (Rydb.) Ewan. Plains Larkspur.
  ssp. *wootoni* x *D. scaposum.*

MYOSURUS. Mousetail.

Myosurus aristatus Benth.
  ssp. *montanus* (Campbell) Stone.
M. cupulatus Wats. Mousetail.
M. minimus L.
M. nitidus Eastw.

RANUNCULUS. Buttercup, Crowfoot.

Ranunculus aquatilis L.
  var. *capillaceus* DC.
R. arizonicus Lemmon.
R. cardiophyllus Hook.
R. cardiophyllus Hook.
  var. *subsagittatus* (Gray) L. Benson.
R. circinatus Sibth.
  var. *subrigidus* (W. Drew) L. Benson.
R. cymbalaria Pursh
  var. *saximontanus* Fern. Desert Crowfoot.
R. eschscholtzii Schlecht.
R. eschscholtzii Schlecht.
  var. *eximius* (Greene) L. Benson.
R. flammula L.
  var. *ovalis* (Bigel.) L. Benson.
R. glaberrimus Hook. (25).
R. hydrocharoides Gray.
R. hydrocharoides Gray
  var. *stolonifer* (Hemsl.) L. Benson.
R. inamoenus Greene.
R. inamoenus Greene
  var. *subaffinis* (Gray) L. Benson.
R. juniperinus Jones.
R. longirostris Godr. White Water Crowfoot (25).

RANUNCULUS. Buttercup, Crowfoot. *(cont.)*
R. *macounii* Britt.
R. *macranthus* Scheele. Large Buttercup.
R. *oreogenes* Greene.
R. *pedatifidus* J. E. Smith
    var. *affinis* (R. Br.) L. Benson.
R. *pensylvanicus* L. f.
R. *scleratus* L.
    var. *multifidus* Nutt.
*R. *testiculatus* Cranz (34).

THALICTRUM. Meadow Rue.

Thalictrum dasycarpum Fisch. & Lall. Purple Meadow Rue.
T. *dasycarpum* Fisch. & Lall.
    var. *hypoglaucum* (Rydb.) Boivin.
T. *fendleri* Engelm.
T. *fendleri* Engelm.
    var. *wrightii* (Gray) Trel.

TRAUTVETTERIA.

Trautvetteria grandis Nutt.

BERBERIDACEAE. Barberry Family.

BERBERIS. Barberry, Mahonia.

Berberis fremontii Torr. Desert Barberry.
B. *haematocarpa* Woot. Red Barberry.
B. *harrisoniana* K. & P.
B. *repens* Lindl. Creeping Barberry.
B. *trifoliata* Moric. Agarito, Algeritas, Currant-of-Texas.
B. *wilcoxii* Kearney.

MENISPERMACEAE. Moonseed Family.

COCCULUS. Snail Seed, Coralbead.

Cocculus diversifolius DC. Correhuela.

PAPAVERACEAE. Poppy Family.

ARCTOMECON. Desert Poppy.

Arctomecon californica Torr. & Frém.
A. *humilis* Coville.

ARGEMONE. Prickly Poppy.

Argemone arizonica G. B. Ownbey.
A. *corymbosa* Greene
    ssp. *arenicola* G. B. Ownbey.
A. *gracilenta* Greene. Crested Pricklepoppy.
A. *mexicana* L. Mexican Poppy, Cardo Santo, Chicalote.
A. *munita* Dur. & Hilg.
    ssp. *argentea* G. B. Ownbey.
    ssp. *rotundata* (Rydb.) G. B. Ownbey.
A. *pleiacantha* Greene. Bluestem Pricklepoppy.
A. *pleiacantha* Greene
    ssp. *ambigua* G. B. Ownbey.

ESCHSCHOLTZIA. California poppy.

Eschscholtzia glyptosperma Greene. Desert Gold Poppy.
E. *mexicana* Greene. Mexican Gold Poppy, Amapola del Campo.
E. *minutiflora* Wats. Little Gold Poppy.

PLATYSTEMON. Cream Cups.

Platystemon californicus Benth.

FUMARIACEAE. Fumitory Family.

CORYDALIS. Scrambled Eggs.

Corydalis aurea Willd. Golden Corydalis.
C. *aurea* Willd.
    occidentalis (Engelm.) G. B. Ownbey.

FUMARIA. Fumitory.

Fumaria occidentalis L. (34).
F. *parviflora* Lam.

CRUCIFERAE. Mustard Family.

ARABIS. Rock Cress.

Arabis drummondii Gray.
A. *fendleri* Greene.
A. *glabra* (L.) Bernh.
A. *gracilipes* Greene.
A. *hirsuta* (L). Scop.
    var. *pycnocarpa* (Hopkins) Rollins.
A. *lignifera* A. Nels.
A. *pendulina* Greene.

ARABIS. Rock Cress. *(cont.)*

A. perennans Wats.

A. pulchra Jones
   var. *pallens* Jones. Prince's Rock Grass.

A. tricornuta Rollins.

ATHYSANUS.

Athysanus pusillus (Hook.) Greene.

A. pusillus (Hook.) Greene
   var. *glabior* Wats.

BARBAREA. Winter Cress.

Barbarea orthoceras Ledeb. American Winter Cress.

B. *orthoceras* Ledeb.
   var. *dolichocarpa* Fern.

BRASSICA. Mustard.

Brassica campestris L. Field Mustard.

B. *hirta Moench*. White Mustard.

B. *juncea* (L.) Cosson. Indian Mustard.

B. *kaber* (DC.) L. C. Wheeler. Charlock.

B. *nigra* (L.) Koch. Black Mustard.

B. *tournefortii* Gouan.

CAMELINA. False Flax.

Camelina microcarpa Andrz. Little Pod.

C. sativa (L.) Crantz. Gold of Pleasure.

CAPSELLA. Shepherds Purse.

Capsella bursa-pastoris (L.) Medic. Paniquesillo.

CARDAMINE. Bitter Cress.

Cardamine cordifolia Gray.

C. parviflora L.

CARDARIA. Hoary Cress.

Cardaria draba (L.) Desv. White Top.

CAULANTHUS.

Caulanthus cooperi (Wats.) Payson (*Thelypodium cooperi* Wats.)
   (50, 97).

C. crassicaulis (Torr.) Wats. Squaw Cabbage.

CHORISPORA.

Chorispora tenella (Pall.) DC.

CONRINGIA. Hares Ear Mustard.

Conringia orientalis (L.) Dum. Treacle Mustard.

CORONOPUS. Wart Cress.

Coronopus didymus (L.) Sm. (94).

DESCURAINIA. Tansy Mustard.

Descurainia californica (Gray) O. E. Schulz.
D. richardsonii (Sweet) O. E. Schulz
D. obtusa (Greene) O. E. Schulz
    ssp. *adenophora* (Woot. & Standl.) Detling.
    ssp. *brevisiliqua* Detling.
D. pinnata (Walt.) Britt. Yellow Tansy Mustard.
D. pinnata (Walt.) Britt.
    ssp. *glabra* (Woot. & Standl.) Detling.
    ssp *halictorum* (Ckll.) Detling.
    ssp. *ochroleuca* (Woot.) Detling.
    ssp. *paysoni* Detling.
D. richardsonii (Sweet) O. E. Schulz
    ssp. *incisa* (Engelm.) Detling.
    ssp. *viscosa* (Rydb.) Detling.
D. sophia (L.) Webb. Flixweed.

DIPLOTAXIS.

Diplotaxis muralis (L.) DC. Stinking Wall Rocket, Sand Rocket.
D. tenuifolia (L.) DC.

DITHYREA. Spectacle Pod.

Dithyrea californica Harv.
D. wislizeni Engelm.

DRABA.

Draba asprella Greene.
D. asprella Greene
    var. *kaibabensis* C. L. Hitchc.
    var. *stelligera* O. E. Schulz.
D. aurea Vahl.
D. aurea Vahl
    var. *leiocarpa* (Payson & St. John) C. L. Hitchc.
D. brachycarpa Nutt.
D. crassifolia Graham.
D. cuneifolia Nutt. Whitlow Grass.

DRABA. *(cont.)*
  D. cuneifolia Nutt.
      var. *integrifolia* Wats.
      var. *platycarpa* (T. & G.) Wats.
  D. helleriana Greene
      var. *bifurcata* C. L. Hitchc.
      var. *blumeri* C .L. Hitchc.
      var. *patens* (Heller) O. E. Schulz.
  D. petrophila Greene.
  D. petrophila Greene
      var. *viridis* (Heller) C. L. Hitchc.
  D. rectifructa C. L. Hitchc.
  D. reptans (Lam.) Fern.
      var. *stellifera* (O. E. Schulz) C. L. Hitchc.
      var. *typica* C. L. Hitchc. *forma micrantha* (Nutt.) C. L. Hitchc.
  D. spectabilis Greene.
  D. standleyi Macbr. & Payson.

DRYOPETALON.
Dryopetalon runcinatum Gray.

ERUCA. Garden Rocket, Roquette.
Eruca sativa Mill.

ERYSIMUM. Wallflower.
Erysimum capitatum (Dougl.) Greene.
E. inconspicuum (Wats.) MacMillan.
E. repandum L.
E. wheeleri Rothr.

HALIMOLOBOS.
Halimolobos diffusus (Gray) O. E. Schulz.

HUTCHINSIA.
Hutchinsia procumbens (L.) Desv.

LEPIDIUM. Pepper Grass, Pepperwort.
Lepidium campestre (L.) R. Br.
L. densiflorum Schrad.
L. densiflorum Schrad.
      var. *bourgeauanum* (Thell.) C. L. Hitchc.
L. fremontii Wats. Desert Alyssum.
L. lasiocarpum Nutt. Sand Peppergrass.
      var. *typicum.*

var. *georginum* (Rydb.) C. L. Hitchc.
var. *wrightii* (Gray) C. L. Hitchc.
L. medium Greene.
L. medium Greene
    var. *pubescens* (Greene) Robins.
L. montanum Nutt.
    var. *alyssoides* (Gray) Jones.
    var. *canescens* (Thell.) C. L. Hitchc.
    var. *glabrum* C. L. Hitchc.
    var. *integrifolium* (Nutt.) C. L. Hitchc.
    var. *jonesii* (Rydb.) C. L. Hitchc.
L. oblongum Small.
L. perfoliatum L. Shield Cress.
L. thurberi Woot.

LESQUERELLA. Bladder Pod.

Lesquerella arizonica Wats.
L. arizonica Wats.
    var. *nudicaulis* Payson.
L. cinerea Wats.
L. fendleri (Gray) Wats.
L. gordoni (Gray) Wats. Gordon Bladderpod.
L. intermedia (Wats.) Heller.
L. ludoviciana (Nutt.) Wats.
L. pinetorum Woot. & Standl. (95).
L. purpurea (Gray) Wats.
L. rectipes Woot. & Standl.
L. tenella Nelson (*L. gordoni* (Gray) Wats. var. *sessilis* Wats.) (95).
L. wardii Wats.

LOBULARIA. Sweet Alyssum.

Lobularia maritima (L.) Desv. (96).

LYROCARPA.

Lyrocarpa coulteri Hook. & Harv. Coulter Lyre Fruit.
    var. *typica* Rollins.

MALCOMIA.

Malcomia africana (L.) R. Br.

MATTHIOLA. Stock.

Matthiola bicornis (Sibth. & Smith) DC. Evening Stock.

PENNELLIA (97).

Pennellia longifolia (Benth.) Roll. (*Thelypodium longifolia* Benth.).
P. micrantha (Gray) Nieuw. (*Thelypodium micranthum* (Gray) Wats.).

PHYSARIA. Twin Pod, Double Bladderpod.

Physaria chambersii Rollins.
P. newberryi Gray.

RAPHANUS.

Raphanus raphanistrum L. Jointed Charlock (13).
R. sativus L. Radish, Rabano (13).

RORIPPA. Yellow Cress.

Rorippa curvisiliqua (Hook.) Bessey.
R. hispida (Desv.) Britt.
R. islandica (Oeder) Borbas. Bog Marsh Cress.
R. nasturtium-aquaticum (L.) Schinz & Thell. Water Cress.
R. obtusa (Nutt.) Britt.
R. sinuata (Nutt.) A. S. Hitchc.
R. sphaerocarpa (Gray) Britt.
R. sylvestris (L.) Besser. Yellow Cress.

SISYMBRIUM.

Sisymbrium altissimum L. Tumble Mustard.
S. ambiguum (Wats.) Payson.
S. irio L. London Rocket.
S. kearneyi Rollins.
S. linifolium Nutt.
S. orientale L. (56).

STANLEYA. Desert Plume, Prince's Plume.

Stanleya albescens Jones.
S. elata Jones.
S. pinnata (Pursh) Britt. Desert Plume.

STREPTANTHELLA.

Streptanthella longirostris (Wats.) Rydb. Long-beaked Twist Flower
S. longirostris (Wats.) Rydb.
   var. *derelicta* J. T. Howell.

STREPTANTHUS. Twist Flower.

Streptanthus arizonicus Wats.
S. *arizonicus* Wats.
  var. *luteus* K. & P.
S. *carinatus* Wright.
S. *cordatus* Nutt.
S. *lemmoni* Wats.

THELYPODIOPSIS

Thelypodiopsis linearifolia (Gray) Al-Shehbaz (*Sisymbrium linearifolium* (Gray) Payson) (97).

THELYPODIUM (97).

Thelypodium integrifolium (Nutt.) Endl.
 ssp. *gracilipes* (Robins.) Al-Shehbaz.
 ssp. *longicarpum* Al-Shehbaz.
T. *lasiophyllum* (H. & A.) Greene.
T. *wrightii* Gray.

THLASPI. Penny Cress (98).

Thlaspi arvense L. Fanweed, Frenchweed, Penny Cress.
T. *montanum* L.
 var. *fendleri* (Gray) P. Holmgren (*T. fendleri* Gray).
 Wild Candytuft.
 var. *montanum*.

THYSANOCARPUS. Lace Pod, Fringe Pod.

Thysanocarpus curvipes Hook.
 var. *elegans* (F. & M.) Robins (*T. amplectans* Greene).
 Fringe Pod (99).
T. *laciniatus* Nutt.
T. *laciniatus* Nutt.
 var. *crenatus* (Nutt.) Brewer.

CAPPARIDACEAE. Caper Family.

ATAMISQUEA.

Atamisquea emarginata Miers.

CLEOMACEAE. Cleome Family.

CLEOME. Spider Flower, Bee Plant (100).
Cleome lutea Hook. Yellow Bee Plant, Yellow Spiderwort.
C. *lutea* Hook.
 var. *jonesii* Macbr. (*C. jonesii* (Macbr.) Tidestrom).

CLEOME. Spider Flower, Bee Plant (100). *(cont.)*

°*C. multicaulis* DC.

C. serrulata Pursh. Rocky Mountain Bee Plant.

C. serrulata Pursh
    forma albiflora (Ckll.) Ckll.

C. sonorae Gray.

CLEOMELLA.

Cleomella longipes Torr.

C. obtusifolia Torr. & Frém. Mohave Stinkweed, Blunt-leaved
Stinkweed.

POLANISIA. Clammy Weed.

Polanisia dodecandra (L.) DC.
    ssp. *trachysperma* (T. & G.) Iltis (P. *trachysperma* T. & G.).
    Western Clammyweed (101).

WISLIZENIA. Jackass Clover.

Wislizenia refracta Engelm. Spectacle Pod.

W. refracta Engelm.
    var. *melilotoides* (Greene) Johnst.

RESEDACEAE. Mignonette Family.

OLIGOMERIS.

Oligomeris linifolia (Vahl) Macbr. Linear-leaved Cambess.

CRASSULACEAE. Orpine Family.

DUDLEYA (50).

Dudleya pulverulenta (Nutt.) B. & R.
    ssp. *arizonica* (Rose) Moran (*Echeveria pulverulenta* Nutt. ssp.
    arizonica (Rose) Clokey).

D. saxosa B. & R.
    ssp. *collomiae* (Rose) Moran (*Echeveria collomiae* (Rose)
    K.&P.) Rock Echeveria.

GRAPTOPETALUM (50).

Graptopetalum bartramii Rose (*Echevaria bartramii* (Rose) K. & P.).

G. rusbyi (Greene) Rose (*Echeveria rusbyi* (Greene) Nels. & Macbr.).

SEDUM. Stonecrop, Orpine.

Sedum cockerellii Britt.
S. *griffithsii* Rose.
S. *lanceolatum* Torr. (*S. stenopetalum* Pursh) (102).
S. *rhodanthum* Gray
S. *stelliforme* Wats.

TILLAEA. Pigmy Weed.

Tillaea aquatica L. Water Pigmy Weed (82).
T. *erecta* H. & A.

SAXIFRAGACEAE. Saxifrage Family.

FENDLERA.

Fendlera rupicola Gray. Fendlerbush.
F. *rupicola* Gray
    var. *falcata* (Thornber) Rehd.
    var. *wrightii* Gray.

FENDLERELLA.

Fendlerella utahensis (Wats.) Heller.
F. *utahensis* (Wats.) Heller
    var. *cymosa* (Greene) K. & P.

HEUCHERA. Alum Root.

Heuchera eastwoodiae Rosendahl et al.
H. *glomerulata* Rosendahl et al.
H. *novomexicana* Wheelock.
H. *parvifolia* Nutt.
    var. *arizonica* Rosendahl et al.
    var. *flavescens* (Rydb.) Rosendahl et al.
H. *rubescens* Torr.
H. *sanguinea* Engelm. Coral Bells.
H. *sanguinea* Engelm.
    var. *pulchra* (Rydb.) Rosendahl.
H. *versicolor* Greene.
H. *versicolor* Greene
    var. *leptomeria* (Greene) K. & P.
    forma pumila Rosendahl et al.

JAMESIA. Cliff Bush.

Jamesia americana T. & G.

LITHOPHRAGMA. Woodland Star.

Lithophragma tenellum Nutt.

PARNASSIA. Grass of Parnassus.

Parnassia parviflora DC.

PHILADELPHUS. Mock Orange, Syringa.

Philadelphus argenteus Rydb.
P. crinitus (C. L. Hitchc.) S. Y. Hu.
P. maculatus (C. L. Hitchc.) S. Y. Hu.
P. madrensis Hemsl.
P. microphyllus Gray.
P. microphyllus Gray
    var. *linearis* S. Y. Hu.
    var. *minutus* (Rydb.) S. Y. Hu.
    var. *ovatus* S. Y. Hu.
P. occidentalis A. Nels.
P. palmeri Rydb.
**P. serpyllifolia Gray* (30).

RIBES. Currant, Gooseberry.

Ribes aureum Pursh. Golden Currant.
R. cereum Dougl. Squaw Currant, Wax Currant.
R. inebrians Lindl. Squaw Currant, Whitestem Gooseberry.
R. inerme Rydb. Whitestem Gooseberry.
R. leptanthum Gray. Trumpet Gooseberry.
R. montigenum McClatchie. Gooseberry Currant.
R. pinetorum Greene. Orange Gooseberry.
R. quercetorum Greene. Oak-belt Gooseberry.
R. velutinum Greene.
R. viscosissimum Pursh. Sticky Currant.
R. wolfii Rothrock.

SAXIFRAGA. Saxifrage.

Saxifraga arguta D. Don. Brook Saxifrage.
S. caespitosa L.
    ssp. *exaratoides* (Simmons) Engl. & Irmsch.
S. debilis Engelm. Pigmy Saxifrage.
S. eriophora Wats.
S. flagellaris Willd.
S. rhomboidea Greene.
S. rhomboidea Greene
    var. *franciscana* (Small) K. & P.

PLANTANACEAE. Plane Tree Family.

PLATANUS. Sycamore, Button Wood, Plane Tree.

Platanus wrightii Wats. Arizona Sycamore.

CROSSOSOMATACEAE. Crossosoma Family.

APACHERIA.

Apacheria chiricahuensis C. T. Mason, Jr. (104).

CROSSOSOMA.

Crossosoma bigelovii Wats. Bigelow Ragged Rock Flower.
C. bigelovii Wats.
    var. *glaucum* (Small) K. & P.
C. parviflorum Robins. & Fern.

ROSACEAE. Rose Family.

AGRIMONIA. Agrimony.

Agrimonia gryposepala Wallr.
A. striata Michx.

AMELANCHIER. Service Berry (103).

Amelanchier pumila Nutt. (*A. polycarpa* Greene).
A. utahensis Koehne (*A. bakeri* Greene, *A. goldmanii*
    Woot. & Standl., *A. mormonica* Schneid., *A. oreophila* A. Nels.).

CERCOCARPUS. Mountain Mahogany (105).

Cercocarpus betuloides Nutt. Birch-leaf Mountain Mahogany.
C. intricatus Wats. Little-leaf Mountain Mahogany.
C. intricatus Wats.
    var. *villosus* C. K. Schneid.
C. ledifolius Nutt.
    var. *intercedens* C. K. Schneid.
C. montanus Raf.
    var. *montanus*. Alder-leaf Mountain Mahogany.
    var. *paucidentatus* (S. Wats.) F. L. Martin (*C. breviflorus*
    Gray and its var. *eximius* C. K. Schneid.).

CHAMAEBATIARIA. Fern Bush, Desert Sweet.

Chamaebatiaria millefolium (Torr.) Maxim.

COLEOGYNE. Black Brush.

Coleogyne ramosissima Torr.

COWANIA. Cliff Rose.

Cowania mexicana D. Don. Quinine Bush.
C. mexicana D. Don
    var. *stansburiana* (Torr.) Jeps.
C. subintegra Kearney.

CRATAEGUS. Hawthorn, Red Haw.

Crataegus erythropoda Ashe. Cerro Hawthorn, Manzanade
    Puyalarga.
C. rivularis Nutt. River Hawthorn.

FALLUGIA. Apache Plume.

Fallugia paradoxa (D. Don) Endl. Ponil.

FRAGARIA. Strawberry.

Fragaria bracteata Heller.
F. ovalis (Lehm.) Rydb.

GEUM. Avens.

Geum macrophyllum Willd.
    var. *perincisum* (Rydb.) Raup. Big-leaf Avens.
G. strictum Ait.
    var. *decurrens* (Rydb.) K. & P.
G. triflorum Pursh
    var. *ciliatum* (Pursh) Fassett. Old Man Whiskers, Grandfather's
    Beard.
G. turbinatum Pursh.

HOLODISCUS. Rock Spiraea.

Holodiscus dumosus (Nutt.) Heller. Shrubby Cream Bush.
H. dumosus (Nutt.) Heller
    var. *australis* (Heller) Ley.

PETROPHYTUM. Rock Mat.

Petrophytum caespitosum (Nutt.) Rydb.
P. caespitosum (Nutt.) Rydb.
    var. *elatius* (Wats.) Tidestrom.

PHYSOCARPUS. Ninebark.

Physocarpus monogynus (Torr.) Coult. Mountain Ninebark.

POTENTILLA. Cinquefoil, Five Finger.

Potentilla albiflora L.
P. anserina L. Silverleaf, Silverweed.

P. anserina L.

    var. *concolor* Ser.

P. arguta Pursh

    ssp. *convallaria* (Rydb.) Keck.

P. biennis Greene.

P. concinna Richards.

P. crinita Gray.

P. crinita Gray

    var. *lemmoni* (Wats.) K. & P.

P. diversifolia Lehm.

P. fruticosa L. Shrubby Cinquefoil.

P. glandulosa Lindl.

    ssp. *arizonica* (Rydb.) Keck.

P. hippiana Lehm.

P. hippiana Lehm.

    var. *diffusa* Lehm.

P. multifoliata (Torr.) K. & P.

P. norvegica L. Rough Cinquefoil.

P. osterhoutii (A. Nels.) J. T. Howell.

P. pensylvanica L.

P. plattensis Nutt.

P. pulcherrima Lehm.

P. rivalis Nutt. Brook Cinquefoil.

P. rivalis Nutt.

    var. *millegrana* (Engelm.) Wats.

P. sibbaldi Hall f.

P. subviscosa Greene.

P. subviscosa Greene

    var. *ramulosa* (Rydb.) K. & P.

P. thurberi Gray.

P. thurberi Gray

    var. *atrorubens* (Rydb.) K. & P.

    var. *sanguinea* (Rydb.) K. & P.

P. viscidula Rydb.

PRUNUS. Plum, Cherry, Peach.

Prunus emarginata (Dougl.) D. Dietr. Bitter Cherry.

P. emarginata (Dougl.) D. Dietr.

    var. *crenulata* (Greene) K. & P.

P. fasciculata (Torr.) Gray. Desert Range Almond.

P. serotina Ehrh. ssp. *virens* (Woot & Standl.) McVaugh.

    Southwestern Black Cherry.

    var. *rufula* (Woot. & Standl.) McVaugh.

    var. *virens.*

PRUNUS. Plum, Cherry, Peach. *(cont.)*

Prunus virginiana L.
    var. *demissa* (Nutt.) Torr. Western Choke Cherry.
    var. *melanocarpa* (A. Nels.) Sarg. Black Western Choke Cherry.

PURSHIA. Antelope Brush.

Purshia tridentata (Pursh) DC.

ROSA. Rose.

Rosa arizonica Rydb.
R. arizonica Rydb.
    var. *granulifera* (Rydb.) K. & P.
R. fendleri Crepin.
R. neomexicana Ckll.
R. stellata Woot. Desert Rose.

RUBUS. Blackberry.

Rubus arizonensis Focke. Arizona Dewberry.
R. leucodermis Dougl. Western Raspberry.
R. neomexicanus Gray. New Mexican Raspberry.
R. parviflorus Nutt.
    var. *parvifolius* (Gray) Fern. Thimbleberry.
R. procerus P. J. Muell. Himalaya Berry.
R. strigosus Michx.
    var. *arizonicus* (Greene) K. & P. American Red Raspberry.

SANGUISORBA. Burnet.

Sanguisorba annua Nutt. Prairie Burnet.
S. minor Scop.

SORBUS. Mountain Ash.

Sorbus dumosa Greene.

VAUQUELINIA. Arizona Rosewood.

Vauquelinia californica (Torr.) Sarg.
*V. pauciflora Standl. (106).

LEGUMINOSAE. Pea Family.
MIMOSOIDEAE. Mimosa Subfamily (107).

ACACIA. Acacia.

Acacia angustissima (Mill.) Kuntze
    var. *shrevi* (B. & R.) Isely (*A. angustissima* (Mill.) *Kuntze* ssp.

lemmonii (Rose) Wiggins). White Ball Acacia.
  var. *suffrutescens* (Rose) Isely (*A. angustissima* (Mill.) Kuntze
  var. *hirta* (Nutt.) Robins.). Fern Acacia.
    var. *texensis* (T. & G.) Isely (*A. angustissima* (Mill.) Kuntze
    var. *cuspidata* L. Benson).
A. *constricta* Benth. White Thorn.
A. *constricta* Benth.
    var. *paucispina* Woot. & Standl.
A. *greggii* Gray
    var. *arizonica* Isely (*A. greggii Gray*). Catclaw.
A. *millefolia* Wats.
A. *neovernicosa* Isely (*A. vernicosa* Standl.).
A. *smallii* Isely (*A. farnesiana* (L.) Willd.). Desert Acacia, Huisache.

CALLIANDRA. False Mesquite.

Calliandra eriophylla Benth. Fairy Duster.
C. *humilis* Benth.
C. *humilis* Benth.
    var. *reticulata* (Gray) L. Benson (*C. reticulata* Gray).
C. *schottii* Torr.

DESMANTHUS. Bundleflower.

Desmanthus cooleyi (Eaton) Trel.
D. *covillei* (B. & R.) Wiggins.
D. *virgatus* (L.) Willd.
    var. *glandulosus* B. L. Turner.

LYSILOMA

Lysiloma microphylla Benth.
    var. *thornberi* (B. & R.) Isely (*L. thornberi* B. & R.).
    Feather Bush, Fern-of-the-Desert.

MIMOSA. Mimosa.

Mimosa biuncifera (Benth.) B. & R. Wait-a-Minute, Cat's Claw
    Mimosa.
M. *dysocarpa* Benth. Gatuno.
M. *dysocarpa* Benth.
    var. *wrightii* (Gray) K. & P.
M. *grahamii* Gray.
M. *grahamii* Gray
    var. *lemmonii* (Gray) K. & P.
M. *laxiflora* Benth.

71

PROSOPIS. Mesquite.

Prosopis articulata S. Wats (?) (108).

P. glandulosa Torr. (*P. juliflora* (Swartz) DC. var. *glandulosa* (Torr.)
Ckll.). Honey Mesquite.

P. glandulosa Torr.
   var. *torreyana* (Benson) M. C. Johnst. (*P. juliflora* (Swartz) DC.
   var. *torreyana Benson*). Western Honey Mesquite.

P. pubescens Benth. Tornillo, Screwbean Mesquite.

P. velutina Woot. (*P. juliflora* (Swartz) DC. var. *velutina* (Woot.)
Sarg.). Velvet Mesquite.

CAESALPINIOIDEAE. Senna Subfamily.

CAESALPINIA.

Caesalpinia gilliesii Wall. Bird of Paradise Flower.

C. jamesii (T. & G.) Fisher (*Hoffmanseggia jamesii* T. & G.) (109).

CASSIA. Senna.

Cassia absus L.

C. armata Wats. Desert Cassia.

C. bauhinioides Gray. Two-leaf Desert Senna.

C. bauhinioides Gray
   var. *arizonica* Robins.

C. covesii Gray. Desert Senna.

C. leptadenia Greenm.

C. leptocarpa Benth. Slim Pod Senna.

C. leptocarpa Benth.
   var. *glaberrima* Jones.

C. lindheimeriana Scheele.

C. wislizenii Gray.

C. wrightii Gray. Patridge Pea.

CERCIDIUM. Palo Verde.

Cercidium floridum Benth. Blue Palo Verde.

C. microphyllum (Torr.) Rose & Johnst. Foothill Palo Verde.

CERCIS. Redbud, Judas Tree.

Cercis occidentalis Torr. California Redbud.

HOFFMANSEGGIA. Rush Pea.

Hoffmanseggia drepanocarpa Gray. Sicklepod Rush Pea.

H. glauca (Ort.) Eifort (*H. densiflora* Benth.). Hog Potato,
   Camote-de-Raton (109).

H. microphylla Torr. Small-leaved Hoffmanseggia.

PARKINSONIA. Jerusalem Thorn.

Parkinsonia aculeata L. Retama, Mexican Palo Verde.

PAPILIONOIDEAE. Bean Subfamily.

AESCHYNOMENE. Joint Vetch.

Aeschynomene villosa Poir.

ALHAGI. Camel Thorn.

Alhagi camelorum Fisch.

AMORPHA. False Indigo.

Amorpha californica Nutt. Stinking Willow.
A. fruticosa L.
    var. *occidentalis* (Abrams) K. & P. Bastard Indigo.

ASTRAGALUS. Milk Vetch, Loco Weed (110).

°Astragalus acutirostris Wats. Keel Beak. First published report
    of this addition to the Arizona Flora.
A. albulus Woot. & Standl.
A. allochrous Gray. Halfmoon Loco.
A. amphioxys Gray.
A. amphioxys Gray
    var. *modestus* Barneby.
    var. *vespertinus* (Sheldon) Jones.
A. argophyllus Nutt.
    var. *martini* Jones (*A. argophyllus* Nutt. var. *pephragmenoides*
    Barneby).
    var. *panguicensis* (Jones) Jones.
A. aridus Gray.
A. arizonicus Gray.
A. atwoodii Welsh & Thorne (111).
A. beathii C. L. Porter.
A. bisulcatus (Hook.) Gray.
A. bisulcatus (Hook.) Gray
    var. *haydenianus* (Gray) Jones.
A. brandegei Porter.
A. bryantii Barneby.
A. calycosus Torr. Gray Locoweed.
A. calycosus Torr.
    var. *scapiosus* (Gray) Jones.
A. castaneiformis Wats.
A. ceramicus Sheldon.
A. cobrensis Gray.

73

ASTRAGALUS. Milk Vetch, Locoweed (110). *(cont.)*
 A. *cobrensis* Gray
    var. *maguirei* Kearney.
 A. *coccineus* Bdge. Scarlet Locoweed.
 A. *coltoni* Jones.
    var. *moabensis* Jones (*A. canovirens* (Rydb.) Barneby).
 A. *crassicarpus* Nutt.
    var. *cavus* Barneby. Ground Plum.
 A. *cremnophylax* Barneby.
 A. *crotalariae* (Benth.) Gray. Desert Rattle Pod.
 A. *desparatus* Jones.
 A. *desparatus* Jones
    var. *conspectus* Barneby.
 A. *didymocarpus* H. & A.
    var. *dispermus* (Gray) Jeps.
 A. *egglestonii* (Rydb.) K. &. P.
 A. *emoryanus* (Rydb.) Cory.
 A. *ensiformis* Jones.
 A. *episcopus* Wats. (*A. kaibensis* Jones).
 A. *eremiticus* Sheldon.
 A. *flavus* Nutt.
 A. *flavus* Nutt.
    var. *candicans* Gray.
 A. *flexuosus* Dougl.
    var. *diehlii* (Jones) Barneby.
    var. *greenei* (Gray) Barneby.
 A. *fucatus* Barneby.
 A. *geyeri* Gray
    var. *triquetrus* (Gray) Jones (*A. triquetrus* Gray).
 A. *gilensis* Greene.
 A. *hallii* Gray
    var. *fallax* (Wats.) Barneby.
 A. *humistratus* Gray.
 A. *humistratus* Gray
    var. *crispulus* Barneby.
    var. *hosackiae* (Greene) Jones.
    var. *humivagans* (Rydb.) Barneby.
    var. *sonorae* (Gray) Jones.
 A. *hypoxylus* Wats.
 A. *insularis* Kell.
    var. *harwoodii* Munz & McBurney. Sand Flat Locoweed.
 A. *kentrophyta* Gray
    var. *coloradensis* Jones.
    var. *elatus* Wats.

A. lancearius Gray.
A. layneae Greene. Layne Locoweed.
A. lentiginosus Dougl.
    var. *ambiguus* Ripley & Barneby.
    var. *australis* Barneby.
    var. *borreganus* Jones.
    var. *diphysus* (Gray) Jones.
    var. *maricopae* Barneby.
    var. *oropedii* Barneby.
    var. *palans* (Jones) Jones.
    var. *stramineus* (Rydb.) Barneby.
    var. *vitreus* Barneby.
    var. *wilsonii* (Greene) Barneby.
    var. *yuccanus* Jones.
A. lonchocarpus Torr.
A. moenocoppensis Jones.
A. mokiacensis Gray.
A. mollissimus Torr.
    var. *bigelovii* (Gray) Barneby (*A. bigelovii* Gray).
    var. *mogollonicus* (Greene) Barneby.
    var. *thompsonae* (Wats.) Barneby (*A. thompsonae* Wats.).
A. monumentalis Barneby.
A. newberryi Gray.
A. newberryi Gray
    var. *blyae* (Rose) Barneby.
A. nothoxys Gray.
A. nuttallianus DC.
    var. *austrinus* (Small) Barneby. Nuttall Locoweed.
    var. *imperfectus* (Rydb.) Barneby.
    var. *micranthiformis* Barneby.
A. oophorus Wats.
    var. *caulescens* (Jones) Jones. Big-podded Locoweed.
A. pattersoni Gray.
A. pinonis Jones.
A. praelongus Sheldon.
A. praelongus Sheldon
    var. *lonshopus* Barneby.
A. preussii Gray.
A. preussii Gray
    var. *laxiflorus* Gray.
A. recurvus Greene.
A. rusbyi Greene.
A. sabulonum Gray.

ASTRAGALUS. Milk Vetch, Locoweed (110). *(cont.)*

A. scopulorum Porter.
A. sesquiflorus Wats.
A. sophoroides Jones.
A. straturensis Jones.
A. subcinereus Gray.
A. tephrodes Gray.
A. tephrodes Gray
    var. *brachylobus* (Gray) Barneby.
    var. *chloridae* (Jones) Barneby.
A. tetrapterus Gray.
A. thurberi Gray.
A. titanophilus Barneby.
A. troglodytus Wats.
A. vaccarum Gray.
A. wingatanus Wats.
A. wootoni Sheldon. Wooton Loco.
A. xiphoides (Barneby) Barneby.
A. zionis Jones.

CLITORIA. Butterfly Pea.

Clitoria mariana L.

COLOGANIA.

Cologania angustifolia H. B. K. (*C. longifolia* Gray) (112).
C. lemmoni Gray.

COURSETIA.

Coursetia microphylla Gray.

CRACCA.

Cracca edwardsii Gray.
C. edwardsii Gray
    var. *glabella* Gray.

CROTALARIA. Rattle Box.

Crotalaria pumila Ort.
C. sagittalis L.
    var. *blumeriana* Senn.
    var. *fruticosa* (Mill.) Fawc. & Rend.

DALEA. Indigo Bush, Pea Bush.

Dalea albiflora Gray.
D. albiflora Gray
 ssp. *villosa* (Rydb.) Wiggins.
D. amoena Wats.
D. amoena Wats.
 var. *pubescens* (Parish) Peebles.
D. aurea Nutt. Golden Dalea.
D. brachystachys Gray.
D. calycosa Gray.
D. diffusa Moric. Escoba Colorado.
D. emoryi Gray. Emory Dalea.
D. filiformis Gray.
D. formosa Torr. Feather Plume.
D. fremontii Torr. Fremont Dalea.
D. fremontii Torr.
 var. *minutifolia* (Parish) L. Benson.
D. grayi (Vail) L. O. Williams.
D. greggii Gray.
D. jamesii (Torr.) T. & G.
D. lachnostachys Gray.
D. lagopus (Cav.) Willd.
D. lemmoni Parry.
D. leporina (Ait.) K. & P. Foxtail Dalea.
D. lumholtzii Robins. & Fern.
D. mollis Benth. Silk Dalea.
D. nana Torr. Dwarf Dalea.
D. nana Torr.
 var. *carnescens* (Rydb.) K. & P.
D. neomexicana (Gray) Cory.
D. neomexicana (Gray) Cory
 ssp. *mollissima* (Rydb.) Wiggins.
D. ordiae Gray.
D. parryi T. & G. Parry Dalea.
D. pogonathera Gray. Herba del Corazon, Bearded Dalea.
D. polygonoides Gray.
D. polygonoides Gray
 var. *anomala* (Jones) Morton.
D. pringlei Gray.
D. schottii Torr. Indigo Bush.
D. scoparia Gray. Broom Pea.
D. spinosa Gray. Smoke Tree.
D. tentaculoides Gentry.
D. terminalis Jones.

DALEA. Indigo Bush, Pea Bush. *(cont.)*

D. thompsonae (Vail) L. O. Williams.

D. urceolata Greene.

D. whitingi K. & P.

D. wislizeni Gray
   ssp. *sessilis* (Gray) Gentry. Indigo Bush.

D. wrightii Gray.

DESMODIUM. Tick Clover, Tick Trefoil.

Desmodium angustifolium (H.B.K.) DC.
   var. *gramineum* (Gray) Schubert.

D. arizonicum Wats.

D. batocaulon Gray.

D. cinerascens Gray.

D. grahami Gray.

D. intortum (Mill.) Urban.

D. metcalfei (Rose & Painter) K. & P.

D. neomexicanum Gray.

D. procumbens (Mill.) A. S. Hitchc.
   var. *exiguum* (Gray) Schubert.

D. psilocarpum Gray.

D. psilophyllum Schlect.

D. retinens Schlect.

D. rosei Schubert.

D. scopulorum Wats.

DIPHYSA.

Diphysa thurberi (Gray) Rydb.

ERRAZURIZIA.

Errazurizia rotundata (Woot.) Barneby (*Parryella rotundata*
   Woot.) (113).

ERYTHRINA. Coral Tree.

Erythrina flabelliformis Kearney. Southwestern Coralbean.

EYSENHARDTIA. Kidney Wood.

° *Eysenhardtia orthocarpa* (Gray) Wats. (50).

E. polystachya (Ort.) Sarg.

GALACTIA.

Galactia wrightii Gray.

G. wrightii Gray
   var. *mollissima* K. &. P.

78

GLYCERRHIZA. Licorice.

Glycyrrhiza lepidota (Nutt.) Pursh.

HEDYSARUM. Sweet Vetch.

Hedysarum boreale Nutt.

INDIGOFERA. Indigo.

Indigofera sphaerocarpa Gray.

LATHYRUS. Peavine.

Lathyrus arizonicus Britt.
L. eucosmus Butters & St. John.
L. graminifolius (Wats.) White.
L. latifolius L. Perennial Sweet Pea.
L. leucanthus Rydb.
    var. *laetivirens* (Green ex Rydb.) C. L. Hitchc.
L. pauciflorus Fern.
    var. *utahensis* (Jones) Peck.
L. zionis C. L. Hitchc.

LOTUS. Deer Vetch, Bird's Foot Trefoil.

Lotus alamosanus (Rose) Gentry.
L. corniculatus L. Birdsfoot Trefoil.
L. greenei (Woot. & Standl.) Ottley.
L. greenei (Woot. & Standl.) Ottley x *L. oroboides* (H.B.K.) Ottley.
L. hamatus Greene.
L. humistratus Greene. Hill Locust.
L. mearnsii Britt.
L. neomexicanus Greene.
L. oblongifolius (Benth.) Greene.
L. oroboides (H.B.K.) Ottley.
L. purshianus (Benth.) Clements & Clements. Spanish Clover.
L. rigidus (Benth.) Greene. Desert Rock Pea.
L. salsuginosus Greene
    var. *brevivexillus* Ottley.
L. tomentellus Greene. Hairy Lotus.
L. utahensis Ottley.
L. wrightii (Gray) Greene. Wright Lotus.

LUPINUS. Lupine.

**Lupinus x alpestris* A. Nels (*L. caudatus x L. argenteus*) (114).
L. argenteus Pursh.
L. arizonicus Wats. Arizona Lupine.
L. barbiger Wats.

LUPINUS. Lupine. *(cont.)*

L. bicolor Lindl.
>var. *microphyllus* (Wats.) C. P. Smith. Pygmy-leaved Lupine.
>var. *pipersmithii* (Heller) C. P. Smith.

L. brevicaulus Wats. Short-stemmed Lupine.

L. caudatus Kell. Tail-cup Lupine, Silver Lupine.

L. caudatus Kell.
>ssp. *cutleri* (Eastw.) Hess & Dunn (*L. cutleri* Eastw.) (114).

L. concinnus Agardh. Elegant Lupine (115).

L. hillii Greene.

L. huachucanus Jones.

L. kingii Wats. Kings Lupine.

L. lemmonii C. P. Smith.

L. neomexicanus Greene.

L. odoratus Heller
>var. *pilosellus* C. P. Smith. Royal Desert Lupine.

L. osterhoutianus C. P. Smith.

L. palmeri Wats.

L. parishii Eastw.

L. pusillus Pursh
>ssp. *intermontanus* (Heller) Dunn. Low Lupine.
>ssp. *rubens* (Rydb.) Dunn.

L. shockleyi Wats. Shockley Lupine.

L. sitgreavesii Wats.

L. sparsiflorus Benth.

L. sparsiflorus Benth.
>ssp. *mohavensis* Dziekanowski & Dunn (115).

L. succulentus Dougl.

L. volutans Greene.

MEDICAGO. Medick.

Medicago hispida Gaertn. Bur Clover.

M. lupulina L. Black Medick.

M. minima (L.) Grufberg. Small Bur Clover.

M. sativa L. Alfalfa.

MELILOTUS. Sweet Clover.

Melilotus albus Desr. White Sweet Clover.

M. indicus (L.) All. Alfalfilla, Annual Yellow Sweet Clover.

M. officinalis (L.) Lam. Yellow Sweet Clover.

NISSOLIA.

Nissolia schottii (Torr.) Gray.

N. wislizeni Gray.

OLNEYA. Tesota.

Olneya tesota Gray. Desert Ironwood, Palofierro, Palo-de-Hierro.

OXYTROPIS. Locoweed.

Oxytropis lambertii Pursh
    var. *bigelovii* Gray. Lambert Locoweed.
O. oreophila Gray.

PARRYELLA.

Parryella filifolia T. & G.

PETALOSTEMUM. Prairie Clover.

P. exile Gray.
P. flavescens Wats.
P. occidentale (Heller) Fern. (*P. candidum* (Willd.) Michx.
    var. *oligophyllum* (Torr.) Herm.). White Prairie Clover (116).
P. purpureum (Vent.) Rydb.
P. searlsiae Gray. Prairie Clover.

PETERIA.

Peteria scoparia Gray. Camote del Monte.
P. thompsonae Wats.

PHASEOLUS. Bean.

Phaseolus acutifolius Gray
    var. *latifolius* Freeman. Tepary Bean.
    var. *tenuifolius* Gray.
P. angustissimus Gray.
P. angustissimus Gray
    var. *latus* Jones.
P. atropurpureus DC. Purple Bean.
P. grayanus Woot. & Standl.
P. heterophyllus Willd.
P. heterophyllus Willd.
    var. rotundifolius (Gray) Piper.
P. leiospermus T. & G.
P. metcalfei Woot. & Standl.
P. parvulus Greene.
P. ritensis Jones.
P. wrightii Gray.

PSORALEA. Scurf Pea.

Psoralea castorea Wats. Beaver Dam Breadroot.
P. epipsila Barneby.
P. juncea Eastw. (34).
P. lanceolata Pursh. Lemon Weed.
P. mephitica Wats.
P. mephitica Wats.
    var. *retrorsa* (Rydb.) K. & P.
P. tenuiflora Pursh
    var. *bigelovii* (Rydb.) Macbr. Scurvy Pea.

RHYNCHOSIA. Rosary Bean.

Rhynchosia rariflora Standl.
R. texana T. & G.

ROBINIA. Locust.

Robinia neomexicana Gray
    var. *luxurians* Dieck. New Mexican Locust.
    var. *subvelutina* (Rydb.) K. & P.

SESBANIA.

Sesbania macrocarpa Muhl. Colorado River Hemp, Bequilla.

SOPHORA.

Sophora arizonica Wats. Arizona Sophora.
S. formosa K. & P.
S. nuttalliana Turner. White Loco (30, 34).
S. sericea Nutt. Silky Sophora.
S. stenophylla Gray.

SPHAEROPHYSA.

Sphaerophysa salsula (Pall.) DC.

SPHINCTOSPERMUM.

Sphinctospermum constrictum (Wats.) Rose.

STYLOSANTHES. Pencil Flower.

Stylosanthes biflora (L.) B.S.P. (?).

TEPHROSIA.

Tephrosia leiocarpa Gray.
T. tenella Gray.
T. thurberi (Rydb.) C. E. Wood.

THERMOPSIS. Golden Pea.

Thermopsis pinetorum Greene.

TRIFOLIUM. Clover.

Trifolium albopurpureum T. & G.

T. amabile H.B.K.

T. andinum Nutt.

T. arizonicum Greene.

T. dubium Sibth. Small Hop Clover.

T. fendleri Greene.

T. fistulosum Vaughan.

T. gracilentum T. & G. Pin-pointed Clover.

T. hybridum L. Alsike Clover.

T. lacerum Greene.

T. microcephalum Pursh.

T. neurophyllum Greene.

T. pinetorum Greene.

T. pratense L. Red Clover.

T. repens L. White Clover.

T. rusbyi Greene.

T. rydbergii Greene (25).

T. subcaulescens Gray.

T. variegatum Nutt.

VICIA. Vetch.

Vicia americana Muhl. American Vetch.

V. americana Muhl.

    var. *minor* Hook. (*V. americana* Muhl. var. *linearis* (Nutt.)
    Wats.) (117).

    var. *truncata* (Nutt.) Brewer.

V. exigua Nutt.

V. leucophaea Greene.

V. pulchella H.B.K.

V. sativa L. Common Vetch.

V. villosa Roth. Hairy Vetch.

ZORNIA.

Zornia diphylla (L.) Pers.

Z. diphylla (L.) Pers.

    var. *leptophylla* Benth.

KRAMERIACEAE. Ratany Family.

KRAMERIA.

Krameria grayi Rose & Painter. White Ratany.
K. lanceolata Torr. Crameria.
K. parvifolia Benth.
    var. *glandulosa* (Rose & Painter) Macbr.
    var. *imparata* Macbr. Little-leaved Ratany.

GERANIACEAE. Geranium Family.

ERODIUM. Heron Bill.

Erodium cicutarium (L.) L'Her. Filaree, Afilerillo, Alfilaria.
*E. moschatum (L.) L'Her.(96.).
E. texanum Gray. Large-flowered Stork's Bill.

GERANIUM. Crane's Bill.

Geranium caespitosum James.
G. carolinianum L.
G. eremophilum Woot. & Standl.
G. fremontii Torr.
G. lentum Woot. & Standl.
G. parryi (Engelm.) Heller.
G. richardsonii Fisch. & Trautv.
G. wislizeni Wats.

OXALIDACEAE. Wood Sorrel Family.

OXALIS. Wood Sorrel (118, 119).

Oxalis albicans H.B.K.
O. albicans H.B.K.
    ssp. *pilosa* (Nutt.) Eiten (*O. pilosa* Nutt.).
O. alpina (Rose) Knuth (*O. metcalfei* (Small) Knuth).
O. amplifolia (Trel.) Knuth.
O. corniculata L. Jocoyote, Agrite, Creeping Wood Sorrel.
O. decaphylla H.B.K. (*O. grayii* (Rose) Knuth).
*O. pes-caprae L. Bermuda Buttercup. This is the first published
    report of this addition to the flora of Arizona.
O. stricta L. Yellow Wood Sorrel, Chanchaquilla.
*O. violacea L. Violet Wood Sorrel.

LINACEAE. Flax Family.

LINUM. Flax (120).

Linum aristatum Engelm.
L. australe Heller
    var. *australe* (*L. aristatum* Engelm. var. *australe* (Heller) K. & P.).
    var. *glandulosum* Rogers.
L. lewisii Pursh. Blue Flax.
L. neomexicanum Greene.
L. puberulum (Engelm.) Heller. Plains Flax.
L. usitatissimum L. Common Flax.

ZYGOPHYLLACEAE. Caltrop Family.

FAGONIA (50).

Fagonia laevis Standl.
F. longipes Standl.
F. pachyacantha Rydb.

KALLSTROEMIA.

Kallstroemia californica (Wats.) Vail. California Caltrop.
K. californica (Wats.) Vail
    var. *brachystylis* (Vail) K. & P.
K. grandiflora Torr. Orange Caltrop.
K. hirsutissima Vail. Carpetweed.
K. parviflora Norton.

LARREA. Creosote Bush.

Larrea tridentata (DC.) Coville. Greasewood, Hediondilla, Gobernadora.

PEGANUM.

Peganum harmala L.

TRIBULUS. Puncture Vine.

Tribulus terrestris L. Cadillo, Abrojo de Flor.

RUTACEAE. Rue Family.

CHOISYA. Star Leaf, Zorillo.

Choisya arizonica Standl.
C. mollis Standl.

PTELEA. Hop Tree (121).

Ptelea trifoliata L.
> ssp. *angustifolia* (Benth.) V. L. Bailey (*P. angustifolia* Benth.).
> Narrowleaf Hoptree, Cola de Zorrillo.
> var. *cognata* (Greene) K. & P.
> ssp. *pallida* (Greene) V. L. Bailey
> var. *lutescens* (Greene) V. L. Bailey.
> var. *pallida* (*P. pallida* Greene). Pale Hoptree.
> ssp. *polyadenia* (Greene) V. L. Bailey.

THAMNOSMA. Dutchman's Breeches.

Thamnosma montana Torr. & Frém. Turpentine Broom.
T. texana (Gray) Torr.

SIMAROUBACEAE. Simarouba Family.

AILANTHUS. Tree of Heaven.

Ailanthus altissima (Mill.) Swingle. Copal Tree.

CASTELA (HOLACANTHA). Crucifixion Thorn (122).

Castela emoryi (A. Gray) Moran & Felger (*Holacantha emoryi* Gray). Corona-de-Cristo, Rosario.

BURSERACEAE. Torch Wood Family.

BURSERA. Bursera.

Bursera fagaroides (H.B.K.) Engler. Fragrant Bursera.
B. microphylla Gray. Elephant Tree, Torote.

MALPIGHIACEAE. Malpighia Family.

ASPICARPA.

Aspicarpa hirtella Rich.

JANUSIA.

Janusia gracilis Gray.

POLYGALACEAE. Milk Wort Family.

MONNINA.

Monnina wrightii Gray.

POLYGALA. Milk Wort.

Polygala acanthoclada Gray. Thorn Polygala.
P. acanthoclada Gray
    var. *intricata* Eastw.
P. alba Nutt. White Milkwort.
P. alba Nutt.
    var. *suspecta* Wats.
P. barbeyana Chodat.
P. glochidiata H.B.K.
P. hemipterocarpa Gray.
P. longa Blake.
P. macradenia Gray.
P. obscura Benth.
P. orthotricha Blake.
P. piliophora Blake.
P. racemosa Blake.
P. reducta Blake.
P. rusbyi Greene.
P. scoparioides Chodat.
P. subspinosa Wats. Spiny Milkwort.
P. tweedyi Britt.

EUPHORBIACEAE. Spurge Family.

ACALYPHA. Three-Seeded Mercury.

Acalypha indica L.
A. lindheimeri Muell. Arg.
A. lindheimeri Muell. Arg.
    var. *major* Pax & Hoff.
A. neomexicana Muell. Arg. New Mexican Copperleaf.
A. ostryaefolia Riddell. Hornbeam Three-seeded Mercury.
A. pringlei Wats.

ARGYTHAMNIA.

Argythamnia brandegei Millsp.
    var. *intonsa* (I. M. Johnst.) Ingram.
A. clariana Jeps.
A. cyanophylla (Woot & Standl.) Ingram.
A. lanceolata (Benth.) Muell. Arg. Lance-leaved Ditaxis.
A. mercurialina (Nutt.) Muell. Arg.
A. neomexicana Muell. Arg.
A. serrata (Torr.) Muell. Arg. Saw-toothed Ditaxis.

BERNARDIA.

Bernardia incana Morton.

CNIDOSCOLUS.

Cnidoscolus angustidens Torr. Mala-Mujer.

CROTON.

Croton californicus Muell. Arg.
 var. *mohavensis* Ferguson. Desert Croton.
C. ciliato-glandulosus Ort.
C. corymbulosus Engelm. Leather Weed.
C. fruticulosus Engelm. Shrubby Croton, Encinilla, Hierbaloca.
C. lindheimerianus Scheele.
C. longipes Jones (34).
C. monanthogynus Michx. Prairie Tea.
C. sonorae Torr.
C. texensis (Klotzsch) Muell. Arg. Dove Weed.

EREMOCARPUS.

Eremocarpus setigerus (Hook.) Benth. Turkey Mullein (55).

EUPHORBIA. Spurge.

Euphorbia abramsiana L. C. Wheeler.
E. albomarginata T. & G. Rattlesnake Weed.
E. alta Norton.
E. arizonica Engelm.
E. bilobata Engelm.
E. brachycera Engelm.
E. capitellata Engelm.
E. chamaesula Boiss.
E. dentata Michx. Toothed Spurge.
E. dentata Michx.
 var. *cuphosperma* Engelm.
E. eriantha Benth. Desert Poinsettia.
E. exstipulata Engelm.
E. fendleri T. & G. Fendler Spurge.
E. fendleri T. & G.
 var. *chaetocalyx* Boiss.
E. florida Engelm.
E. glyptosperma Englm. Rib-seeded Sand Mat.
E. gracillima Wats.
E. heterophylla L. Painted Spurge, Catalina.
E. heterophylla L.
 var. *graminifolia* (Michx.) Engelm.

88

E. hirta L.

E. hyssopifolia L. Hyssop Spurge.

E. incisa Engelm.

E. incisa Engelm.
  var. *mollis* (Norton) L. C. Wheeler.

E. indivisa (Engelm.) Tidestrom.

E. lathyris L. Myrtle Spurge.

E. lurida Engelm.

E. marginata Pursh. Snow-on-the-Mountain.

E. melanadenia Torr.

E. micromera Boiss. Sonoran Sand Mat.

E. ocellata Dur. & Hilg.
  var. *arenicola* (Parish) Jeps.

E. odontadenia Boiss.

E. palmeri Engelm.

E. palmeri Engelm.
  var. *subpubens* (Engelm.) L. C. Wheeler.

E. parryi Engelm. Parry Euphorbia.

E. pediculifera Engelm.

E. peplus L. Petty Spurge.

E. platysperma Engelm.

E. plummerae Wats.

E. polycarpa Benth. Small-seeded Sand Mat.

E. polycarpa Benth.
  var. *hirtella* Boiss.

E. prostrata Ait. Groundfig Spurge.

E. radians Benth.

E. revoluta Engelm.

E. robusta (Engelm.) Small.

E. serpens H.B.K.

E. serpyllifolia Pers.

E. serrula Engelm. Sawtooth Spurge.

E. setiloba Engelm. Bristle-lobed Sand Mat.

E. spathulata Lam.

E. stictospora Engelm. Narrow-seeded Spurge.

E. supina Raf. Prostrate Spurge.

E. trachysperma Engelm.

E. vermiculata Raf.

JATROPHA.

Jatropha cardiophylla (Torr.) Muell. Arg. Limber Bush.
    Sangre-de-Cristo.
J. cinerea (Ort.) Muell. Arg.
J. cuneata Wiggins & Rollins. Sangre-de-Drago.
J. macrorhiza Benth.
    var. *septemfida* Engelm.

MANIHOT.

Manihot angustiloba (Torr.) Muell. Arg.
M. davisiae Croizat.

REVERCHONIA.

Reverchonia arenaria Gray.

RICINUS. Castor Bean.

Ricinus communis L. Higuerilla.

SAPIUM.

Sapium biloculare (Wats.) Pax. Mexican Jumping Bean.

STILLINGIA.

Stillingia linearifolia Wats.
S. paucidentata Wats.
S. spinulosa Torr. Broad-leaved Stillingia.

TETRACOCCUS.

Tetracoccus fasciculatus (Wats.) Croizat
    var. *hallii* (T. S. Brand.) Dressler.

TRAGIA.

°Tragia amblyodonta (Muell. Arg.) Pax & Hoffm. (123).
T. laciniata (Torr.) Muell. Arg.
T. nepetaefolia Cav.
T. ramosa Torr. (T. stylaris Muell. Arg.) (123).

CALLITRICHACEAE. Water Starwort Family.

CALLITRICHE. Water Starwort.

°Callitriche hermaphroditica L. (25).
C. heterophylla Pursh.
C. verna L.

BUXACEAE. Box Family.

SIMMONDSIA. Jojoba, Deer Nut.

Simmondsia chinensis (Link) Schneid. Goat Nut, Coffee Berry.

ANACARDIACEAE. Cashew Family, Sumac Family.

RHUS. Sumac.

Rhus choriophylla Woot. & Standl. Mearns Sumac.
R. *glabra* L. Smooth Sumac.
R. *kearneyi* Barkley. Kearney Sumac.
R. *microphylla* Engelm. Desert Sumac.
R. *ovata* Wats. Sugar Sumac.
R. *radicans* L.
    var. *rydbergii* (Small) Rehder. Poison Ivy.
R. *trilobata* Nutt. Squaw Bush.
R. *trilobata* Nutt.
    var. *anisophylla* (Greene) Jeps.
    var. *pilosissima* Engler.
    var. *quinata* Jeps.
    var. *racemulosa* (Greene) Barkley.
    var. *simplicifolia* (Greene) Barkley.

CELASTRACEAE. Bitter Sweet Family.

CANOTIA.

Canotia holacantha Torr.

GLOSSOPETALON. Grease Bush.

Glossopetalon nevadense Gray.
G. *spinescens* Gray. Spiny-stemmed Tongue Flower.

MORTONIA.

Mortonia scabrella Gray.
M. *scabrella* Gray
    var. *utahensis* Coville.

PACHYSTIMA. Box Leaf.

Pachystima myrsinites (Pursh) Raf. Mountain Lover, Oregon
    Boxwood.

ACERACEAE. Maple Family.

ACER. Maple.

Acer glabrum Torr. Rocky Mountain Maple.
A. glabrum Torr.
  var. *diffusum* (Greene) Smiley.
  var. *neomexicanum* (Greene) K. & P.
A. grandidentatum Nutt. Big Tooth Maple, Palo de Azucar.
A. grandidentatum Nutt.
  var. *brachypterum* (Woot. & Standl.) E. J. Palmer.
A. negundo L.
  var. *interius* (Britt.) Sarg. Box Elder, Fresno de Guajuco.

SAPINDACEAE. Soapberry Family.

CARDIOSPERMUM.

Cardiospermum halicacabum L. Balloon Vine, Farolitos.

DODONAEA. Hop Bush.

Dodonaea viscosa Jacq.
  var. *angustifolia* (L. f.) Benth.

SAPINDUS. Soapberry.

Sapindus saponaria L.
  var. *drummondii* (H. & A.) Benson. Western Soapberry,
  Jaboncillo.

MELIACEAE. Melia Family.

MELIA. Chinaberry.

Melia azedarach L. Umbrella Tree.

RHAMNACEAE.

CEANOTHUS.

Ceanothus fendleri Gray. Buck Brush, Deer Brier.
C. fendleri Gray
  var. *venosus* Trel.
  var. *viridis* Gray.
C. greggii Gray.

C. greggii Gray
    var. *orbiculatus* Kelso.
    var. *perplexans* (Trel.) Jeps.
C. integerrimus H. & A.
    var. *californicus* (Kell.) Benson. Deer Brush.
C. martini Jones.

COLUBRINA. Snakewood.

Colubrina californica Johnst. California Snake Bush.

CONDALIA (incl. MICRORHAMNUS) (124).

Condalia correllii M. C. Johnst.
C. ericoides (A. Gray) M. C. Johnst. (*Microrhamnus*
    ericoides A. Gray). Javelina Brush, Little Buck Thorn.
C. globosa Johnst.
    var. *pubescens* Johnst. Bitter Condalia.
C. warnockii M. C. Johnst.
    var. *kearneyana* M. C. Johnst. (30).

RHAMNUS. Buck Thorn.

Rhamnus betulaefolia Greene. Birch Leaf Buck Thorn.
R. betulaefolia Greene
    var. *obovata* K. & P.
R. californica Esch.
    ssp. *ursina* (Greene) Wolf. California Buck Thorn.
R. crocea Nutt. Red Berry Buck Thorn.
R. crocea Nutt.
    var. *illicifolia* (Kell.) Greene. Hollyleaf Buck Thorn.
R. serrata Schult. (*R. smithii* Greene ssp. *fasciculata* (Greene)
    C. B. Wolf) (125).

SAGERETIA.

Sageretia wrightii Wats.

ZIZYPHUS.

Zizyphus obtusifolia (Hook. ex. T. & G.) A. Gray
    var. *canescens* (A. Gray) M. C. Johnst. (*Condalia lycioides*
    (Gray) Weberb. var. *canescens* (Gray) Trel.). Thorn,
    Gray-leaved Abrojo (124).

VITACEAE. Grape Family.

CISSUS.

Cissus trifoliata L.

PARTHENOCISSUS. Virginia Creeper.

Parthenocissus inserta (Kerner) K. Fritsch. Thicket Creeper.

VITIS. Grape.

Vitis arizonica Engelm. Canyon Grape, Parra del Monte.
V. arizonica Engelm.
  var. *glabra* Munson.

TILIACEAE. Linden Family.

CORCHORUS.

Corchorus hirtus L. Orinoco Jute, Moralia.

MALVACEAE. Mallow Family.

ABUTILON. Indian Mallow.

Abutilon californicum Benth.
A. incanum (Link) Sweet. Indian Mallow, Pelotazo.
A. palmeri Gray.
A. parishii Wats.
A. parvulum Gray. Small-leaved Abutilon.
A. pringlei Hochr.
A. reventum Wats.
A. sonorae Gray.
A. theophrasti Medic. Velvet Leaf, Chingma.
A. thurberi Gray.

ANODA.

Anoda abutiloides Gray.
A. crenatiflora Ort.
A. cristata (L.) Schlecht. Spurred Anoda.
A. cristata (L.) Schlecht
  var. *digitata* (Gray) Hochr.
A. pentaschista Gray.
A. reticulata Wats.
A. thurberi Gray.
A. wrightii Gray.

CALLIRHOË. Poppy Mallow.

Callirhoë involucrata (T. & G.) Gray. Wine Cup.

EREMALCHE.

Eremalche exilis (Gray) Greene.
E. rotundifolia (Gray) Greene. Desert Five Spot.

GOSSYPIUM. Cotton.

Gossypium thurberi Todaro. Desert Cotton.

HERISSANTIA.

Herissantia crispa (L.) Brizicky (*Bogenhardia crispa* (L.)
Kearney) (126).

HIBISCUS. Rose Mallow.

Hibiscus biseptus Wats.
H. coulteri Harv. Desert Rose Mallow.
H. denudatus Benth. Rock Hibiscus.
H. denudatus Benth.
var. *involucellatus* Gray.

HORSFORDIA.

Horsfordia alata (Wats.) Gray. Pink Felt Plant.
H. newberryi (Wats.) Gray. Yellow Felt Plant.

ILIAMNA. Wild Hollyhock.

Iliamna grandiflora (Rydb.) Wiggins.

MALVA. Mallow, Cheese.

Malva neglecta Wallr. Common Mallow.
M. parviflora L. Little Mallow.
°M. sylvestris L.
ssp. *mauritiana* (L.) Boiss. High Mallow (13).

MALVASTRUM.

Malvastrum bicuspidatum (Wats.) Rose.

MALVELLA (127).

Malvella lepidota (A. Gray) Fryxell (*Sida lepidota* Gray).
Scurfy Sida.
M. leprosa (Ort.) Krapov (*Sida leprosa* (Ort.) K. Schum.).

SIDA.

Sida filicaulis T. & G. (*S. procumbens Sw.*) (128).

S. *hederacea* (Dougl.) Torr. Alkali Sida.

S. *neomexicana* Gray.

S. *physocalyx* Gray. Tuberous Sida.

S. *rhombifolia* L. Axocatzin.

S. *spinosa* L.

    var. *angustifolia* (Lam.) Griseb. Prickly Mallow.

S. *tragiaefolia* Gray.

SIDALCEA. Checker Mallow.

Sidalcea neomexicana Gray. Alkali Pink.

SPHAERALCEA. Alkali Pink.

Sphaeralcea ambigua Gray. Desert Mallow.

S. *ambigua* Gray

    var. *rosacea* (Munz & Johnston) Kearney.

S. *angustifolia* (Cav.) G. Don

    var. *cuspidata* Gray. Narrow-leaved Globe Mallow.

    var. *oblongifolia* (Gray) Shinners (var. *lobata*
    (Wooton) Kearney) (129).

S. *coccinea* (Pursh) Rydb. Scarlet Globe Mallow.

S. *coccinea* (Pursh) Rydb.

    var. *dissecta* (Nutt.) Garrett.

    var. *elata* (Baker f.) Kearney.

S. *coulteri* (Wats.) Gray. Coulter Globe Mallow.

S. *digitata* (Greene) Rydb. Juniper Globe Mallow.

S. *emoryi* Torr. Emory Globe Mallow.

S. *emoryi* Torr.

    var. *arida* (Rose) Kearney.

    var. *californica* (Parish) Shinners (Var. *variabilis*
    (Ckll.) Kearney) (129).

    var. *nevadensis* Kearney.

S. *fendleri* Gray. Fendler Globe Mallow.

S. *fendleri* Gray

    var. *albescens* Kearney.

    var. *elongata* Kearney.

    var. *tripartita* (Woot. & Standl.) Kearney.

    var. *venusta* Kearney.

S. *grossulariaefolia* (H. & A.) Rydb.

S. *grossulariaefolia* (H. & A.) Rydb.

    var. *pedata* (Torr.) Kearney.

S. *incana* Torr.

S. *incana* Torr.
  var. *cuneata* Kearney.

S. *laxa* Woot. & Standl. Caliche Globe Mallow.

S. *leptophylla* (Gray) Rydb. Scaly Globe Mallow.

S. *orcuttii* Rose. Orcutt Globe Mallow.

S. *parvifolia* A. Nels. Littleleaf Globe Mallow.

S. *rusbyi* Gray.

S. *rusbyi* Gray
  var. *gilensis* Kearney.

S. *subhastata* Coult.

S. *subhastata* Coult.
  var. *connata* Kearney.
  var. *pumila* (Woot. & Standl.) Kearney.
  var. *thyrsoidea* Kearney.

S. *wrightii* Gray.

STERCULIACEAE. Cacao Family.

AYENIA.

Ayenia compacta L. (*A. pusilla* L.) (130).
*A. *filiformis* Wats. (30).
A. *microphylla* Gray.

FREMONTODENDRON.

Fremontodendron californicum (Torr.) Coville. Flannel Bush.

HERMANNIA.

Hermannia pauciflora Wats. Hierba del Soldado.

WALTHERIA.

Waltheria americana L.

GUTTIFERAE.

HYPERICUM. St. John's Wort.

Hypericum anagalloides C. & S. Tinker's Penny.
H. *formosum* H.B.K.

ELATINACEAE. Water Wort Family.

ELATINE. Water Wort.

Elatine brachysperma Gray.
*E. californica Gray (131).
*E. chilensis Gray (25).
*E. rubella Rydb. (34).

TAMARICACEAE. Tamarix Family.

TAMARIX. Tamarisk.

Tamarix aphylla (L.) Karst. Athel (34, 50).
T. pentandra Pall. Salt Cedar.

COCHLOSPERMACEAE. Cochlospermum Family.

AMOREUXIA.

Amoreuxia gonzalezii Sprague & Riley.
A. palmatifolia M. & S.

KOEBERLINIACEAE. Junco Family.

KOEBERLINIA. All Thorn.

Koeberlinia spinosa Zucc.
K. spinosa Zucc.
    var. *tenuispina* K. & P.

VIOLACEAE. Violet Family.

HYBANTHUS. Green Violet.

Hybanthus attenuatus (H. & B.) C. K. Schulze.
H. verticillatus (Ort.) Baill.

VIOLA. Violet.

Viola adunca J. E. Smith.
V. canadensis L.
V. canadensis L.
    var. *rydbergii* (Greene) House.
    var. *scariosa* Porter.
V. charlestonensis Baker & Clausen.
V. nephrophylla Greene.
V. nephrophylla Greene
    var. *arizonica* (Greene) K. & P.
V. nuttallii Pursh.

*V. *palustris* L.
    ssp. *brevipes* Baker (133).
  V. *pedatifida* G. Don. Larkspur Violet.
  V. *purpurea* Kell.
  V. *purpurea* Kell.
    ssp. *mohavensis* (Baker) Clausen (V. *aurea* Kell.
    ssp. *arizonensis* Baker & Clausen) (132).
  V. *rafinesquii* Greene. Field Pansy.
  V. *umbraticola* H.B.K.
    var. *glaberrima* Baker.

PASSIFLORACEAE. Passion Flower Family.

PASSIFLORA. Passion Flower.

Passiflora bryonioides H.B.K.
P. *foetida* L.
    var. *arizonica* Killip. White Passion Flower, Corona de Cristo.
P. *mexicana* Juss.

LOASACEAE. Stick Leaf Family.

CEVALLIA.

Cevallia sinuata Lag.

EUCNIDE.

Eucnide urens Parry. Sting Bush.

MENTZELIA. Blazing Star.

Mentzelia affinis Greene. Yellow Comet.
M. *albicaulis* Dougl. Small-flowered Blazing Star.
M. *aspera* L.
M. *involucrata* Wats.
M. *multiflora* (Nutt.) Gray. Adonis Blazing Star.
M. *multiflora* (Nutt.) Gray
    var. *integra* Jones.
M. *nitens* Greene. Venus Blazing Star.
M. *nitens* Greene
    var. *jonesii* (Urban & Gilg.) J. Darl.
    var. *leptocaulis* J. Darl.
M. *puberula* J. Darl. Rough-stemmed Blazing Star.
M. *pumila* (Nutt.) T. &. G.
M. *texana* Urb. & Gilb. (M. *asperula* Woot. & Standl.).
M. *tricuspis* Gray. Spiny-leaved Blazing Star.
M. *veatchiana* Kell.

PETALONYX. Sandpaper Plant.

Petalonyx linearis Greene. Long-leaved Sandpaper Plant.
P. nitidus Wats. Shiny-leaved Sandpaper Plant.
P. parryi Gray.
P. thurberi Gray. Thurber Sandpaper Plant.

SYMPETELEIA.

Sympeteleia rupestris (Baill.) Gray. Flor de la Piedra.

CACTACEAE. Cactus Family (3).

CEREUS.

Cereus giganteus Engelm. (*Carnegiea gigantea* (Engelm.)
B. & R.) Saguaro.
C. greggii Engelm. (*Peniocereus greggii* (Engelm.) B. & R.).
Desert Night-blooming Cereus.
C. greggii Engelm.
var. *transmontanus* Engelm.
C. schottii Engelm. (*Lophocereus schottii* (Engelm.) B. & R.). Senita.
C. striatus Brand. (*Wilcoxia diguetii* (Weber) Peebles).
C. thurberi Engelm. (*Lemaireocereus thurberi* (Engelm.) B. & R.).
Organpipe Cactus, Pitahaya.

COCHISEIA (134).

Cochiseia robbinsorum Earl.

CORYPHANTHA.

Coryphantha missouriensis (Sweet) B. & R.
C. missouriensis (Sweet) B. & R.
var. *marstonii* (Clover) L. Benson.
C. recurvata (Engelm.) B. & R. (*Mammillaria recurvata* Engelm.).
C. scheeri (Kuntze) L. Benson.
C. Scheeri (Kuntze) L. Benson
var. *robustispina* (Schott) L. Benson (*Mammillaria
robustispina* Schott).
var. *valida* (Engelm.) L. Benson.
C. strobiliformis (Poseiger) Orcutt.
C. strobiliformis (Poseiger) Orcutt
var. *orcuttii* (Rose) L. Benson.
C. vivipara (Nutt.) B. & R.

C. vivipara (Nutt.) B. & R.
    var. *alversonii* (Coult.) L. Benson.
    var. *arizonica* (Engelm.) W. T. Marshall (*Mammillaria arizonica* Engelm.).
    var. *bisbeeana* (Orcutt) L. Benson.
    var. *desertii* (Engelm.) W. T. Marshall (*Mammillaria chlorantha* Engelm.).
    var. *rosea* (Clokey) L. Benson.

ECHINOCACTUS.

Echinocactus horizonthalonius Lemaire.
E. horizonthalonius Lemaire
    var. *nicholii* L. Benson.
E. polycephalus Engelm. & Bigel. Niggerhead Cactus.
E. polycephalus Engelm. & Bigel.
    var. *xeranthemoides* Coult. (*E. xeranthemoides* Engelm. ex Rydb.)

ECHINOCEREUS. Hedgehog Cactus, Strawberry Cactus.

Echinocereus engelmannii (Parry) Lemaire.
E. engelmannii (Parry) Lemaire
    var. *acicularis* L. Benson.
    var. *chrysocentrus* (Engelm.) Engelm. ex Rumpler.
    var. *decumbens* L. Benson (?).
    var. *nicholii* L. Benson.
    var. *variegatus* (Engelm.) Engelm. ex Rumpler.
E. fasciculatus (Engelm.) L. Benson (*E. fendleri* Engelm.
    var. *robustus* L. Benson & *Mammillaria fasciculata* Engelm.).
E. fasciculatus (Engelm.) L. Benson
    var. *bonkerae* (Thornber & Bonker) L. Benson
    (*E. boyce-thompsonii* Orcutt var. *bonkarae* Peebles).
    var. *boyce-thompsonii* (Orcutt) L. Benson
    (*E. boyce-thompsonii* Orcutt).
E. fendleri Engelm.
E. fendleri Engelm.
    var. *rectispinus* (Peebles) L. Benson.
E. ledingii Peebles.
E. pectinatus (Scheidw.) Engelm.
E. pectinatus (Scheidw.) Engelm.
    var. *neomexicanus* (Coult.) L. Benson.
    var. *rigidissimus* (Engelm.) Engelm. ex Rumpler.
    Rainbow Cactus.
E. triglochidiatus Engelm.

ECHINOCEREUS. Hedgehog Cactus, Strawberry Cactus. *(cont.)*

E. triglochidiatus Engelm.
    var. *arizonicus* (Rose ex Orcutt) L. Benson.
    var. *gonacanthus* (Engelm. & Bigel.) Boiss.
    var. *melanacanthus* (Engelm.) L. Benson
    (*Mammillaria aggregata* Engelm.).
    var. *mojavensis* (Engelm.) L. Benson.
    var. *neomexicanus* (Standl.) Standl. ex W. T. Marshall
    (E. triglochidiatus Engelm. var. *polycanthus* (Engelm.)
    L. Benson).

EPITHELANTHA. Button Cactus.

Epithelantha micromeris (Engelm.) Weber.

FEROCACTUS. Barrel Cactus, Bisnaga.

Ferocactus acanthodes (Lemaire) B. & R.
F. acanthodes (Lemaire) B. & R.
    var. *eastwoodiae* L. Benson.
    var. *lecontei* (Engelm.) Lindsay (*F. lecontei* (Engelm.) B. & R.).
F. covillei B. & R.
F. wislizenii (Engelm.) B. & R.

MAMMILLARIA. Fishhook Cactus, Pincushion Cactus.

Mammillaria grahamii Engelm.
M. grahamii Engelm.
    var. *oliveiae* (Orcutt) L. Benson (M. *oliviae* Orcutt).
M. gummifera Engelm.
M. gummifera Engelm.
    var. *applanata* (Engelm.) L. Benson (*M. heyderi* Muehlenpfordt
    var. *applanata* Engelm.).
    var. *macdougalii* (Rose) L. Benson (*M. macdougalii* Rose).
    var. *meiacantha* (Engelm.) L. Benson.
M. lasiacantha Engelm.
M. mainiae K. Brand.
M. microcarpa Engelm.
M. orestera L. Benson.
M. tetrancistra Engelm.
M. thornberi Orcutt.
M. viridiflora (B. & R.) Bodeker (135).
M. wrightii Engelm.
    var. *wilcoxii* (Toumey ex K. Schumann) W. T. Marshall
    (*M. wilcoxii* Toumey) (135).
    var. *wrightii.*

NEOLLOYDIA.

Neolloydia erectocentra (Coult.) L. Benson (*Echinomastus erectocentrus* (Coult.) B.&.R.).
N. erectocentra (Coult.) L. Benson
  acunensis (W. T. Marshall) L. Benson.
N. intertexta (Engelm.) L. Benson (*Echinomastus intertextus* (Engelm.) B. & R.).
N. johnsonii (Parry) L. Benson (*Echinomastus johnsonii* (Parry) Baxter and var. *lutescens* Parish).

OPUNTIA.

Opuntia acanthocarpa Engelm. & Bigel. Buckhorn Cholla.
O. acanthocarpa Engelm. & Bigel.
  var. *coloradensis* L. Benson.
  var. *major* (Engelm. & Bigel.) L. Benson (*O. acanthocarpa* Engelm. & Bigel. var. *ramosa* Peebles).
  var. *thornberi* (Thornber & Bonker) L. Benson (*O. thornberi* Thornber & Bonker).
O. arbuscula Engelm. (*O. vivipara* Rose). Pencil Cholla.
O. basilaris Engelm. & Bigel. Beavertail Cactus.
O. basilaris Engelm. & Bigel.
  var. *aurea* (Baxter) W. T. Marshall (*O. aurea* Baxter). Yellow Beavertail.
  var. *longiareolata* (Clover & Jotter) L. Benson. Grand Canyon Beavertail.
  var. *treleasei* (Coulter) Toumey. Kern Cactus.
O. bigelovii Engelm. Teddy Bear Cactus.
O. canada Griffiths (*O. phaeacantha* Engelm.
  var. *laevis* X *major* and *O. gilvescens* Griffiths).
O. chlorotica Engelm. & Bigel. Pancake Pear.
O. clavata Engelm. Club Cholla.
O. curvospina Griffiths (136).
O. echinocarpa Engelm. & Bigel. Silver Cholla, Golden Cholla.
O. erinacea Engelm. & Bigel. Mohave Prickly Pear.
O. erinacea Engelm. & Bigel.
  var. *hystricina* (Engelm. & Bigel.) L. Benson (*O. hystricina* Engelm. & Bigel.). Porcupine Prickly Pear.
  var. *ursina* (Weber) Parish (*O. ursina* Weber). Grizzly Bear Cactus.
  var. utahensis (Engelm.) L. Benson (O. rhodantha Schum.).
O. fragilis Nutt. Little Prickly Pear.
O. fragilis Nutt.
  var. *brachyarthra* (Englm. & Bigel.) Coult.
O. fulgida Engelm. Jumping Cholla.

103

OPUNTIA. *(cont.)*

O. fulgida Engelm.
    var. *mammillata* (Schott) Coult.
O. imbricata Haw. Tree Cholla.
O. kleiniae DC.
    var. *tetracantha* (Toumey) W. T. Marshall
    (*O. tetracantha* Toumey).
O. leptocaulis DC. Desert Christmas Cactus.
O. littoralis (Engelm.) Ckll.
O. macrorhiza Engelm. (*O plumbea* Rose). Plains Prickly Pear.
O. macrorhiza Engelm.
    var. *pottsii* (Salm-Dyck) L. Benson.
O. nicholii L. Benson. Navajo Bridge Prickly Pear.
O. phaeacantha Engelm.
O. phaeacantha Engelm.
    var. *discata* (Griffiths) Benson & Walkington
    (*O. engelmannii* Salm-Dyck). Engelmann Prickly Pear.
    var. *flavispina* L. Benson (137).
    var. *laevis* (Coult.) L. Benson (*O. laevis* Coult.).
    var. *major* Engelm.
    var. *superbospina* (Griffiths) L. Benson (137).
O. polycantha Haw. Plains Prickly Pear.
O. polycantha Haw.
    var. *juniperina* (Engelm.) L. Benson.
    var. *rufispina* (Engelm.) L. Benson.
    var. *trichophora* (Engelm. & Bigel.) Coult.
O. pulchella Engelm. Sand Cholla.
O. ramosissima Engelm. Diamond Cholla.
O. spinosior (Engelm.) Toumey. Cane Cholla.
O. spinosior x O. versicolor (138).
O. stanlyi Engelm. Devil Cholla.
O. stanlyi Engelm.
    var. *kunzei* (Rose) L. Benson (*O. kunzei* Rose *and O. kunzei*
    Rose var. *wrightiana* (Baxter) Peebles). Kunze Cholla.
    var. *parishii* (Orcutt) L. Benson (*O. parishii* Orcutt).
    Parish Cholla.
    var. *peeblesiana* L. Benson.
O. versicolor Engelm. Staghorn Cholla.
O. violacea Engelm. Purple Prickly Pear.

O. violacea Engelm.
 var. *gosseliniana* (Weber) L. Benson.
 var. *macrocentra* (Engelm.) L. Benson (*O. macrocentra*
 Engelm.). Black-spined Prickly Pear.
 var. *santa-rita* (Griffiths & Hare) L. Benson (*O. santa-rita*
 (Griffiths & Hare) Rose).
O. whipplei Engelm. & Bigel. (*O. whipplei* Engelm. & Bigel.
 var. *enodis* Peebles). Whipple Cholla.
O. whipplei Engelm. & Bigel.
 var. *multigeniculata* (Clokey) L. Benson.
O. wigginsii L. Benson.

PEDIOCACTUS.

Pediocactus bradyi L. Benson.
P. papyracanthus (Engelm.) L. Benson (*Toumeya papyracanthus*
 (Engelm.) B. & R.) Grama Grass Cactus.
P. paradinei B. W. Benson.
P. peeblesianus (Croizat) L. Benson (*Navajoa peeblesiana Croizat*).
P. peeblesianus (Croizat) L. Benson
 var. *fickeiseniae* L. Benson.
 var. *maianus* L. Benson.
P. sileri (Engelm.) L. Benson (*Utahia sileri* (Engelm.) B. & R.).
P. simpsonii (Engelm.) B. & R.

SCLEROCACTUS (139).

Sclerocactus parviflorus Clover & Jotter
 var. *intermedius* (Peebles) Woodruff & Benson (*S. whipplei*
 (Engelm. & Bigel.) B. & R. var. *intermedius* (Peebles) L. Benson;
 S. intermedius Peebles).
S. pubispinus (Engelm.) L. Benson
 var. *sileri* H. Benson.
S. spinosior (Engelm.) Woodruff & L. Benson.
S. whipplei (Engelm. & Bigel.) B. & R. (*S. whipplei* (Engelm. &
 Bigel.) B. & R. var. *pygmaeus* Peebles).
S. whipplei (Engelm. & Bigel.) B. & R.
 var. *roseus* (Clover) L. Benson (*S. parviflorus* Clover & Jotter).

ELAEAGNACEAE. Oleaster Family.

ELAEAGNUS. Oleaster.

Elaeagnus angustifolia L. Russian Olive.

SHEPHERDIA. Buffalo Berry.

Shepherdia argentea (Pursh) Nutt. Silver Buffalo Berry.
S. canadensis (L.) Nutt.
S. rotundifolia Parry. Round Leaf Buffalo Berry.

LYTHRACEAE. Loosestrife Family.

AMMANNIA.

Ammannia auriculata Willd.
    var. *arenaria* (H.B.K.) Koehne (?).
A. coccinea Rottb. Tooth Cup.

CUPHEA.

Cuphea wrightii Gray.
C. wrightii Gray
    var. *nematopetala* Bacigalupi.

LYTHRUM. Loosestrife.

Lythrum californicum T. & G. Hierba del Cancer.

ROTALA.

Rotala ramosior (L.) Koehne. Tooth Cup.

ONAGRACEAE. Evening Primrose Family (140).

BOISDUVALIA.

**Boisduvalia densiflora* (Lindl.) S. Wats. (82).
**B. glabella* (Nutt.) Walp. (13).

CALYLOPHUS (141).

Calylophus hartwegii (Benth.) Raven (*Oenothera hartwegii* Benth.).
    ssp. *fendleri* (A. Gray) Towner & Raven (*Oenothera hartwegii*
    Benth. var. *fendleri* (A. Gray) A. Gray).
    ssp. *pubescens* (A. Gray) Towner & Raven
    (*Oenothera greggii* A. Gray).
C. lavandulifolius (Torr. & A. Gray) Raven (*Oenothera*
    lavandulaefolia Torr. & A. Gray and its var. *glandulosa* Munz).
C. serrulatus (Nutt.) Raven (*Oenothera serrulata* Nutt.).
C. toumeyi (Small) Towner (*Oenothera hartwegii* Benth.
    var. *toumeyi* (Small) Munz).

CAMISSONIA (142).

Camissonia arenaria (N. Nels.) Raven (*Oenothera cardiophylla*
(Torr.) Raven var. *splendens* Munz & Johnst.).
C. boothii (Dougl.) Raven (*Oenothera boothii* Dougl.)
Booth Primrose.
C. boothii (Dougl.) Raven
ssp. *condensata* (Munz) Raven (*Oenothera decorticans*
(H. & A.) Greene var. *condensata* Munz).
ssp. *decorticans* (H. & A.) Raven (*Oenothera decorticans*
(H. & A.) Greene).
C. brevipes (Gray) Raven (*Oenothera brevipes* Gray). Yellow Cups.
C. brevipes (Gray) Raven
ssp. *pallidula* (Munz) Raven (*Oenothera pallidula* Munz).
C. cardiophylla (Torr.) Raven (*Oenothera cardiophylla* Torr.)
Heart-leaved Primrose.
C. chamaenerioides (Gray) Raven (*Oenothera chamaenerioides*
Gray). Long-capsuled Primrose.
C. clavaeformis (Torr. & Frém.) Raven
ssp. *aurantiaca* (Wats.) Raven (*Oenothera clavaeformis* Torr.
& Frém. var. *aurantiaca* (Wats.) Munz). Brown-eyed Primrose.
ssp. *peeblesii* (Munz) Raven (*Oenothera clavaeformis*
Torr. & Frém. var. *peeblesii* Munz).
ssp. *Peirsonii* (Munz) Raven (*Oenothera clavaeformis*
Torr. & Frém. var. *peirsonii* Munz).
C. contorta (Dougl.) Kearney (*Oenothera contorta* Dougl. including
var. *epilobioides* (Greene) Munz). Dwarf Contorted Primrose.
C. micrantha (Hornem.) Raven (*Oenothera micrantha* Hornem.).
C. multijuga (Wats.) Raven (*Oenothera multijuga* Wats.).
C. multijuga (Wats.) Raven
ssp. *orientalis* (Munz) Raven (*Oenothera multijuga* Wats.
var. *orientalis* Munz).
ssp. *parviflora* (Wats.) Raven (*Oenothera multijuga* Wats.
var. *parviflora* (Wats.) Munz). Frost-stemmed Primrose.
C. pallida (Abrams) Raven (*Oenothera micrantha* Hornem.
var. *exfoliata* (A. Nels.) Munz). Spencer Primrose.
C. parryi (Wats.) Raven (*Oenothera parryi* Wats.).
C. pterosperma (Wats.) Raven (*Oenothera pterosperma* Wats.)
Pigmy Primrose.
C. refracta (Wats.) Raven (*Oenothera refracta* Wats.).
Narrow-leaved Primrose.
C. scapoidea (Nutt.) Raven (*Oenothera scapoidea* Nutt.
var. *seorsa* (A. Nels.) Munz).
C. speculicola (Raven) Raven.

CIRCAEA. Enchanter's Nightshade.

Circaea alpina L.
  ssp. *pacifica* (Asch. & Magnus) M. E. Jones (*C. pacifica*
  Asch. & Magnus).

CLARKIA.

Clarkia epilobioides (Nutt.) Nels. & Macbr.
C. purpurea (Curt.) Nels. & Macbr.
  ssp. *quadrivulnera* (Dougl.) Lewis & Lewis.
C. rhomboidea Dougl.

EPILOBIUM. Willow Weed (25).

Epilobium adenocaulon Hausskn.
E. adenocaulon Hausskn.
  var. *parishii* (Trel.) Munz.
E. angustifolium L.
  ssp. *circumvagum* Mosquin. Fire Weed, Blooming Sally (143).
E. californicum Hausskn.
E. ciliatum Raf.
E. foliosum (T. & G.) Suksd. (*E. minutum* Lindl. ex Lehm.) (144).
E. halleanum Hausskn.
E. hornemanni Reichenb.
E. oregonense Hausskn.
E. paniculatum Nutt.
  forma adenocladen Hausskn.
  forma subulatum Hausskn.
  forma tracyi (Rydb.) St. John.
E. saximontanum Hausskn.

GAURA (145).

Gaura coccinea Pursh (incl. vars. *arizonica* Munz; *epilobioides*
  (H.B.K.) Munz; *glabra* (Lehm.) T. & G.; *parvifolia* (Torr.)
  T. & G.). Scarlet Gaura.
G. hexandra Ort.
  ssp. *gracilis* (Woot. & Standl.) Raven & Gregory (*G. gracilis*
  Woot. & Standl. and *forma glandulosa* (Woot. & Standl.) Munz).
G. parviflora Dougl. (incl. *forma glabra* Munz and *forma*
  lachnocarpa Weatherby). Lizard Tail, Velvet Leaf Gaura.

GAYOPHYTUM (146).

Gayophytum decipiens Lewis & Szweykowski.
G. diffusum T. & G.
  ssp. *parviflorum* Lewis & Szweykowski (*G. nuttallii* T. & G.
  var. *intermedium* (Rydb.) Munz).
G. ramosissimum T. & G.

LUDWIGIA. Water Primrose, Seed Box (140).

Ludwigia palustris (L.) Ell. (*L. palustris* (L.) Ell. var. *americana* (DC.) Fern. & Griscom). Marsh Purslane.

L. peploides (H.B.K.) Raven (*Jussiaea repens* L. var. *peploides* (H.B.K.) Griseb.). Yellow Water Weed, Verdolaga de Agua.

°L. repens Forst. (*L. natans* Ell.). Water Primrose (82).

OENOTHERA. Evening Primrose, Sun Drops.

Oenothera albicaulis Pursh. Prairie Evening Primrose.

O. brachycarpa Gray
    var. *wrightii* (Gray) Léveillé.

O. caespitosa Nutt.
    var. *australis* (Woot. & Standl.) Munz.
    var. *jonesii* Munz.
    var. *marginata* (Nutt.) Munz. Large White Desert Primrose.
    var. *montana* (Nutt.) Durand.

O. cavernae Munz.

O. coronopifolia T. & G.

O. deltoides Torr. & Frém. Dune Primrose.

O. deltoides Torr. & Frém.
    var. *arizonica* Munz.
    var. *cineracea* (Jeps.) Munz.
    var. *decumbens* (Wats.) Munz.
    var. *piperi* Munz.

O. flava (A. Nels.) Garrett.

O. hookeri T. & G.
    ssp. *hewettii* Ckll.
    ssp. *hirsutissima* (Gray) Munz.

O. kunthiana (Spach) Munz.

O. laciniata Hill
    var. *pubescens (Willd.)* Munz. Cut-leaved Evening Primrose.

O. leptocarpa Greene.

O. longissima Rydb.

O. longissima Rydb.
    ssp. *clutei* (A. Nels.) Munz.

O. neomexicana (Small) Munz.

O. pallida Lindl.

O. pallida Lindl.
    ssp. *runcinata* (Engelm.) Munz & Klein (*O. runcinata* (Engelm.)
    Munz and its var. *brevifolia* (Engelm.) Munz and var.
    leucotricha (Woot. & Standl.) Munz).

°O. platanorum Raven & Parnell (147).

O. primiveris Gray. Large Yellow Desert Primrose.

109

OENOTHERA. Evening Primrose, Sun Drops. *(cont.)*

  O. primiveris Gray
    var. *caulescens* Munz.
  O. procera Woot. & Standl.
  O. rosea Ait. Rose Sundrops.
  O. speciosa Nutt. Amapola del Campo, Showy Primrose.
  O. taraxacoides (Woot. & Standl.) Munz.
  O. trichocalyx Nutt.

ZAUSCHNERIA. Hummingbird Trumpet.

Zauschneria californica Presl
  ssp. *latifolia* (Hook.) Keck.
*Z. *garrettii* Nels. (90).

HIPPURIDACEAE. Mare's Tail Family.

HIPPURIS. Mare's Tail.

Hippuris vulgaris L.

HALORAGACEAE. Water Milfoil Family.

MYRIOPHYLLUM. Water Milfoil.

Myriophyllum brasiliense Cambess. Parrot's Feather, Water
  Feather.
M. spicatum L.
  ssp. *exalbescens* (Fern.) Hult.

ARALIACEAE. Ginseng Family.

ARALIA. Spikenard.

Aralia humilis Cav.
A. racemosa L. American Spikenard.

UMBELLIFERAE. Parsley Family.

ALETES.

Aletes macdougali Coult. & Rose.

AMMI.

*Ammi majus L. Bishop's Weed (148).

AMMOSELINUM. Sand Parsley.

Ammoselinum giganteum Coult. & Rose.

ANETHUM. Dill.

Anethum graveolens L.

ANGELICA.

Angelica pinnata Wats.

ANTHRISCUS.

Anthriscus scandicina (Weber) Mansfield (82).

APIASTRUM.

Apiastrum angustifolium Nutt. Wild Celery.

APIUM. Celery.

Apium graveolens L. Common Celery.
A. leptophyllum (Pers.) F. Muell. Wild Celery.

BERULA. Water Parsnip.

Berula erecta (Huds.) Coville.

BOWLESIA.

Bowlesia incana Ruiz & Pavon. Hairy Bowlesia.

BUPLEURUM. Thorough Wax.

Bupleurum subovatum Link.

CAUCALIS.

Caucalis microcarpa H. & A.

CHAEROPHYLLUM. Chervil.

Chaerophyllum tainturieri Hook. (149).

CICUTA. Water Hemlock.

Cicuta douglasii (DC.) Coult. & Rose.

CONIOSELINUM. Hemlock Parsley.

Conioselinum mexicanum Coult. & Rose.
C. scopulorum (Gray) Coult. & Rose.

CONIUM. Poison Hemlock.

Conium maculatum L.

CORIANDRUM. Coriander.

Coriandrum sativum L.

CYMOPTERUS.

Cymopterus bulbosus A. Nels.
C. fendleri Gray.
C. megacephalus Jones.
C. multinervatus (Coult. & Rose) Tidestrom. Purple Cymopterus.
C. newberryi (Wats.) Jones.
C. purpurascens (Gray) Jones.
C. purpureus Wats.

DAUCUS. Carrot.

*Daucus carota L. Wild Carrot, Queen Anne's Lace (82).
D. pusillus Michx. American Carrot.

ERYNGIUM. Eryngo, Button Snakeroot.

Eryngium heterophyllum Engelm. Mexican Thistle.
E. lemmoni Coult. & Rose.
E. phyteumae Delar.
E. sparganophyllum Hemsl.

FOENICULUM. Fennel.

Foeniculum vulgare Mill.

HERACLEUM. Cow Parsnip.

Heracleum lanatum Michx. Masterwort.

HYDROCOTYLE. Water Pennywort.

Hydrocotyle ranunculoides L. f.
H. verticillata Thunb.
H. verticillata Thunb.
    var. *triradiata* (A. Rich.) Fern.

LIGUSTICUM.

Ligusticum porteri Coult. & Rose. Chuchupate, Osha.

LILAEOPSIS.

Lilaeopsis recurva A. W. Hill.

LOMATIUM.

Lomatium dissectum (Nutt.) Mathias & Constance
    var. *multifidum* (Nutt.) Mathias & Constance.
L. foeniculaceum (Nutt.) Coult. & Rose
    var. *macdougali* (Coult. & Rose) Cronq. (*L. macdougali*
    Coult. & Rose) (150).
L. leptocarpum (T. & G.) Coult. & Rose.

L. nevadense (Wats.) Coult. & Rose
  var. *parishii* (Coult. & Rose) Jeps. Parish Wild Parsley.
  var. *pseudorientale* (Jones) Munz.
L. parryi (Wats.) Macbr. Parry Wild Parsley.

OREOXIS.

Oreoxis alpina (Gray) Coult. & Rose.

OSMORHIZA. Sweet Cicely.

Osmorhiza brachypoda Torr.
O. chilensis H. & A.
O. depauperata Phil.

OXYPOLIS. Hog Fennel.

Oxypolis fendleri (Gray) Heller.

PASTINACA. Parsnip.

Pastinaca sativa L.

PERIDERIDIA (151).

Perideridia parishii (Coult. & Rose) Nels. & Macbr.
  ssp. *parishii* Chuang & Constance.

PSEUDOCYMOPTERUS.

Pseudocymopterus montanus (Gray) Coult. & Rose. Mountain
  Parsley.

PTERYXIA.

Pteryxia davidsoni (Coult. & Rose) Mathias & Constance.
P. petraea (Jones) Coult. & Rose.

SCANDIX. Venus' Comb.

Scandix pecten-veneris L. Shepherd's Needle, Crow Needles.

SIUM. Water Parsnip.

Sium suave Walt.

SPERMOLEPIS. Scale Seed.

Spermolepis echinata (Nutt.) Heller.

TORILIS. Hedge Parsley.

Torilis nodosa (L.) Gaertn. Knotted Hedge Parsley.

CORNACEAE. Dogwood Family.

CORNUS. Dogwood, Cornel.
Cornus stolonifera Michx. Red Osier Dogwood.

GARRYACEAE. Silk Tassel Family.

GARRYA. Silk Tassel.
Garrya flavescens Wats. Quinine Bush.
G. wrightii Torr.

LENNOACEAE. Lennoa Family.

AMMOBROMA.
Ammobroma sonorae Torr. Sand Sponge, Biatatk.

PHOLISMA.
Pholisma arenarium Nutt. Scaly-stemmed Sand Plant.

ERICACEAE. Heather Family.

ARBUTUS. Madrono, Madrone.
Arbutus arizonica (Gray) Sarg. Arizona Madrone.

ARCTOSTAPHYLOS. Manzanita.
Arctostaphylos patula Greene. Green Leaf Manzanita.
A. pringlei Parry.
A. pungens H.B.K. Mexican Manzanita.
A. uva-ursi (L.) Spreng.

MONOTROPA. Pinesap.
Monotropa latisquama (Rydb.) Hultén.

PHYLLODOCE. Mountain Heath.
Phyllodoce empetriformis (J. E. Smith) D. Don.

PTEROSPORA. Pine Drops.
Pterospora andromeda Nutt.

VACCINIUM. Blueberry.
Vaccineum oreophilum Rydb.

PYROLACEAE. Pyrola Family.

CHIMAPHILA. Pipsissewa.

Chimaphila maculata (L.) Pursh
    var. *dasystemma* (Torr.) K. & P.
C. umbellata (L.) Nutt.
    var. *acuta* (Rydb.) Blake.

MONESES.

Moneses uniflora (L.) Gray.

PYROLA. Wintergreen, Shinleaf.

Pyrola elliptica Nutt.
P. picta J. E. Smith.
P. picta J. E. Smith
    forma aphylla (J. E. Smith) Camp.
P. secunda L. (*Ramischia secunda* (L.) Garcke). Side-bells Pyrola.
P. virens Schweigg.

PRIMULACEAE. Primrose Family.

ANAGALLIS. Pimpernel.

Anagallis arvensis L. Scarlet Pimpernel, Herba de Pajaro.
A. arvensis L.
    forma caerulea (Schreb.) Baumg. Poorman's Weather
    Glass (96).

ANDROSACE. Rock Jasmine.

Androsace occidentalis Pursh.
A. occidentalis Pursh
    var. *arizonica* (Gray) St. John.
A. septentrionalis L.
    var. *glandulosa* (Woot. & Standl.) St. John.
    var. *puberulenta* (Rydb.) Knuth.
    var. *subulifera* Gray.

CENTUNCULUS. Chaff Weed.

Centunculus minimum L. False Pimpernel.

DODECATHEON. Shooting Star.

Dodecatheon alpinum (Gray) Greene
    ssp. *majus* H. J. Thompson.
D. dentatum Hook.
    ssp. *ellisiae* (Standl.) H. J. Thompson.
D. pulchellum (Raf.) Merrill.

LYSIMACHIA. Loosestrife.

Lysimachia ciliata L.
  var. *validula* (Greene) K. & P. Fringed Loosestrife.

PRIMULA. Primrose.

Primula hunnewellii Fern.
P. parryi Gray.
P. rusbyi Greene.
P. specuicola Rydb.

SAMOLUS.

Samolus parviflorus Raf. (*S. floribundus* H.B.K.).
S. vagans Greene.

PLUMBAGINACEAE. Plumbago Family.

LIMONIUM. Sea Lavender, Marsh Rosemary.
*Limonium limbatum Small (13, 131).

PLUMBAGO. Leadwort.

Plumbago scandens L. Hierba de Alacran, Pitillo.

FOUQUIERIACEAE. Ocotillo Family.

FOUQUIERIA. Ocotillo, Candlewood.

Fouquieria splendens Engelm. Coach Whip.

SAPOTACEAE. Sapote Family.

BUMELIA.

Bumelia lanuginosa (Michx.) Pers.
  var. *rigida* Gray. Gum Bumelia.

OLEACEAE. Olive Family.

FORESTIERA. Adelia.

Forestiera neomexicana Gray. Tanglebrush.
F. neomexicana Gray
  var. *arizonica* Gray. Desert Olive.
F. shrevei Standl.

FRAXINUS. Ash.

Fraxinus anomala Torr. Single-leaved Ash.
F. cuspidata Torr.
    var. *macropetala* (Eastw.) Rehd. Fragrant Ash.
F. dipetala H. & A. Two Petal Ash.
F. gooddingii Little. Goodding's Ash.
F. papillosa Lingelsheim. Chihuahua Ash.
F. pennsylvanica Marsh.
    ssp. *velutina* (Torr.) G. N. Miller. Velvet Ash.

MENODORA.

Menodora scabra Gray.
M. scabra Gray
    var. *laevis* (Woot. & Standl.) Steyerm.
    var. *longituba* Steyerm.
    var. *ramosissima* Steyerm.
M. scoparia Engelm. Broom Twinberry.
M. spinescens Gray. Spiny Mendora.

LOGANIACEAE. Logania Family.

BUDDLEJA. Butterfly Bush.

Buddleja scordioides H.B.K. Escobilla (148).
B. sessiliflora H.B.K. Tepozan.
B. utahensis Coville.

GENTIANACEAE. Gentian Family.

CENTAURIUM. Canchalagua, Centaury.

Centaurium calycosum (Buckl.) Fern. Buckley's Centaury.
C. nudicaule (Engelm.) Robins.

EUSTOMA.

Eustoma exaltatum (L.) Griseb. Catchfly Gentian.

GENTIANA. Gentian.

Gentiana affinis Griseb. Pleated Gentian.
G. bigelovii Gray.
G. farwoodii Gray.
G. fremontii Torr. Moss Gentian.
G. parryi Engelm.

GENTIANELLA.

Gentianella amarella (L.) Börner
    ssp. *acuta* (Michx.) J. M. Gillett (*Gentiana strictiflora* (Rydb.) A. Nels.).
    ssp. *heterosepala* (Engelm.) J. M. Gillett (*Gentiana heterosepala* Engelm.).
    ssp. *wrightii* (Gray) J. M. Gillett (*Gentiana wrightii* Gray).
G. barbellata (Engelm.) J. M. Gillett (*Gentiana barbellata* Engelm.).
G. detonsa (Rottb.) G. Don
    ssp. *elegans* (A. Nels.) J. M. Gillett (*Gentiana thermalis* Kuntze). Western Fringed Gentian.
    ssp. *superba* (Greene) J. M. Gillett (*Gentiana grandis* (Gray) Holm).
G. microcalyx (Lemmon) J. M. Gillett (*Gentiana microcalyx* Lemmon).
G. tenella (Rottb.) Börner (*Gentiana tenella* Rottb.).
G. wislizeni (Engelm.) J. M. Gillett (*Gentiana wislizeni* Engelm.).

HALENIA. Spurred Gentian.

Halenia recurva (J. E. Smith) Allen.

MENYANTHES.

Menyanthes trifoliata L.

NYMPHOIDES.

Nymphoides peltatum (Gmel.) Britten & Rendle. Yellow Floating Heart (82).

SWERTIA. Elkweed, Green Gentian.

Swertia albomarginata (Wats.) Kuntze.
S. perennis L. Felwort.
S. radiata (Kell.) Kuntze. Deers Ears.
S. radiata (Kell.) Kuntze
    var. *macrophylla* (Greene) St. John.
S. utahensis (Jones) St. John.

APOCYNACEAE. Dogbane Family.

AMSONIA. Blue Star.

Amsonia brevifolia Gray (152).
A. eastwoodiana Rydb.
A. grandiflora Alexander.
A. hirtella Standl.
A. hirtella Standl.
    var. *pogonosepala* Wiggins.

A. jonesii Woodson.
A. kearneyana Woodson.
A. palmeri Gray.
A. peeblesii Woodson.
A. tomentosa Torr. & Frém.
A. tomentosa Torr. & Frém.
    var. *stenophylla* K. & P.

APOCYNUM. Dogbane.

Apocynum androsaemifolium L. Spreading Dogbane.
A. androsaemifolium L.
    var. *glabrum* Macoun.
A. cannabinum L. Hemp, Dogbane, Indian Hemp.
A. cannabinum L.
    var. *glaberrimum* A. DC.
A. jonesii Woodson.
A. medium Greene.
A. medium Greene
    var. *floribundum* (Greene) Woodson.
A. sibiricum Jacq.
    var. *salignum* (Greene) Fern. Clasping Leaf Dogbane.
A. suksdorfii Greene.
A. suksdorfii Greene
    var. *angustifolium* (Wooton) Woodson. Prairie Dogbane.

HAPLOPHYTON.

Haplophyton crooksii L. Cockroach Plant, Hierba-de-la-Cucuracha.

MACROSIPHONIA. Rock Trumpet.

Macrosiphonia brachysiphon (Torr.) Gray.

ASCLEPIADACEAE. Milkweed Family (153).

ASCLEPIAS. Milkweed, Silkweed.

Asclepias albicans Wats. White-stemmed Milkweed.
A. angustifolia Schweig.
A. asperula (Decne.) Woodson
    ssp. *capricornu* (Woodson) Woodson (*A. capricornu*
    Woodson) (30).
A. brachystephana Engelm. Short-crowned Milkweed.
A. cryptoceras Wats.
A. cutleri Woodson.
A. elata Benth.
A. engelmanniana Woodson.

119

ASCLEPIAS. Milkweed, Silkweed. *(cont.)*

A. *erosa* Torr. Desert Milkweed.

A. *hallii* Gray.

A. *hypoleuca* (Gray) Woodson.

A. *involucrata* Engelm.

A. *latifolia* (Torr.) Raf. Broadleaf Milkweed.

A. *lemmoni* Gray.

A. *linaria* Cav.

A. *macrosperma* Eastw.

A. *macrotis* Torr.

A. *nummularia* Torr. Yerba de Cuerve.

A. *nyctaginifolia* Gray. Four O'Clock Milkweed.

A. *quinquedentata* Gray.

A. *rusbyi* (Vail) Woodson.

A. *speciosa* Torr. Showy Milkweed.

A. *subulata* Decne. Desert Milkweed, Ajamete.

A. *subverticillata* (Gray) Vail. Poison Milkweed, Western Whorled Milkweed.

A. *tuberosa* L.
    ssp. *interior* Woodson. Butterfly Weed, Pleurisy Root, Orange Milkweed.

A. *uncialis* Greene.

A. *viridiflora* Raf.
    var. *lanceolata* (Ives) Torr.

CYNANCHUM.

Cynanchum arizonicum (Gray) Shinners (*Metastelma arizonicum* Gray).

C. *sinaloense* Woods. (*Mellichampia sinaloensis* (T.S.Bdge.) K. & P.).

C. *utahense* (Engelm.) Woods. (*Astephanus utahensis* Engelm.). Deboltia.

*C. *wigginsii* Shinners (*Basistelma angustifolium* (Torr.) Bartlett).

MATELEA.

Matelea arizonica (A. Gray) Shinners (*Lachnostoma arizonicum* Gray).

M. *balbisii* (Decne.) Woods. (*Pherotrichis balbisii* (Decne.) Gray).

M. *cordifolia* (Gray) Woods. (*Rothrockia cordifolia* Gray).

M. *parvifolia* Torr.) Woods. (*Gonolobus parvifolius* Torr.).

M. *producta* (Torr.) Woods. (*Gonolobus productus* Torr.).

SARCOSTEMMA. Climbing Milkweed.

Sarcostemma crispum Benth. (*Funastrum crispum* (Benth.)
Schlechter).
S. *cynanchoides* Decne. (*Funastrum cynanchoides* (Decne.)
Schlechter). Climbing Milkweed.
S. *hirtellum* (Gray) Holm (*Funastrum hirtellum* (Gray) Schlechter).
Rambling Milkweed.

CONVOLVULACEAE. Morning Glory Family.

CALYSTEGIA. Hedge Bindweed (154).

Calystegia sepium (L.) R. Br. (*Convolvulus sepium* L.).
C. *sepium* (L.) R. Br.
    var. *fraterniflora* (Mack. & Bush) Shinners (*Convolvulus
sepium* L. var. *fraterniflora* Mack. & Bush) (25).

CONVOLVULUS. Bindweed.

Convolvulus arvensis L. Field Bindweed.
C. *equitans* Benth. (*C. incanus* Vahl). Hoary Bindweed.
C. *linearilobus* Eastw.

CRESSA. Alkali Weed.

Cressa truxillensis H.B.K.

CUSCUTA. Dodder.

Cuscuta applanata Engelm.
C. *californica* H. & A. California Dodder.
C. *campestris* Yuncker.
C. *cephalanthi* Engelm. Buttonbush Dodder (25).
C. *coryli* Engelm. Hazel Dodder.
C. *dentatasquamata* Yuncker.
C. *denticulata* Engelm. Toothed Dodder.
C. *erosa* Yuncker.
C. *gronovii* Willd.
C. *indecora* Choisy. Pretty Dodder.
C. *mitraeformis* Engelm.
C. *odontolepis* Engelm.
C. *salina* Engelm.
C. *tuberculata* T. S. Brand.
C. *umbellata* H.B.K. Umbrella Dodder.
C. *umbellata* H.B.K.
    var. *reflexa* (Coult.) Yuncker.
°C. *warneri* Yuncker (25).

DICHONDRA. Pony Foot.

Dichondra argentea Willd. Silver Pony Foot.
D. brachypoda Woot. & Standl. New Mexican Pony Foot.
D. repens Forst.
    var. *sericea* (Swartz) Choisy.

EVOLVULUS.

Evolvulus alsinoides L.
    var. *acapulcensis* (Willd.) Van Ooststroom. Dio de Vibora.
E. arizonicus Gray.
E. arizonicus Gray
    var. *laetus* (Gray) Van Ooststroom.
E. pilosus Nutt.
E. sericeus Swartz.
E. sericeus Swartz
    var. *discolor* (Benth.) Gray.

IPOMOEA. Morning Glory.

Ipomoea barbatisepala Gray.
I. coccinea L. Scarlet Creeper, Scarlet Morning Glory.
I. costellata Torr.
* I. cristulata* Hallier f. (30).
I. egregia House.
* I. hederacea* (L.) Jacq. (25, 34).
I. heterophylla Ort.
I. hirsutula Jacq. f. Woolly Morning Glory.
I. lemmoni Gray.
I. leptotoma Torr.
I. leptotoma Torr.
    var. *wootoni* Kelso.
I. longifolia Benth.
I. muricata Cav.
I. plummerae Gray.
I. purpurea (L.) Roth. Tall Morning Glory.
I. tenuiloba Torr.
I. thurberi Gray.
I. triloba L.

JACQUEMONTIA.

Jacquemontia palmeri Wats.
J. pringlei Gray.

POLEMONIACEAE. Phlox Family.

ALLOPHYLLUM.

Allophyllum gilioides (Benth.) A. & V. Grant (*Gilia gilioides* (Benth.) Greene). Straggling Gilia (155).

COLLOMIA.

Collomia grandiflora Dougl.
C. linearis Nutt.

ERIASTRUM.

Eriastrum diffusum (Gray) Mason.
E. diffusum (Gray) Mason
    ssp. *jonesii* Mason.
E. eremicum (Jeps.) Mason.
E. eremicum (Jeps.) Mason
    ssp. *yageri* (Jones) Mason.

GILIA.

Gilia achilleaefolia Benth.
    ssp. *multicaulis* (Benth.) V. & A. Grant.
    ssp. *staminea* (Greene) Mason & Grant.
G. clokeyi Mason.
G. filiformis Parry. Thread-stemmed Gilia.
G. flavocincta A. Nels.
G. gunnisoni T. & G.
G. hutchinsifolia Rydb.
G. latifolia Wats. Broad-leaved Gilia.
G. laxiflora (Coult.) Osterh.
G. leptomeria Gray. Tooth-leaved Gilia.
G. mexicana A. & V. Grant.
G. minor A. & V. Grant.
G. ophthalmoides Brand.
G. rigidula Benth.
    var. *acerosa* Gray.
G. scopulorum Jones. Rock Gilia.
G. sinuata Dougl.
G. stellata Heller.
G. subnuda Torr.
    ssp. *superba* (Eastw.) Brand.
G. transmontana (Mason & A. Grant) A. & V. Grant.

IPOMOPSIS (156).

Ipomopsis aggregata (Pursh) V. Grant (*Gilia aggregata* (Pursh) Spreng.). Sky Rocket.

I. aggregata (Pursh) V. Grant
    ssp. *arizonica* (Greene) V. & A. Grant (*Gilia aggregata* (Pursh) Spreng. var. *arizonica* (Greene) Fosberg).

I. longiflora (Torr.) V. Grant (*Gilia longiflora* (Torr.) G. Don). White-flowered Gilia.

I. macombii (Torr.) V. Grant (*Gilia macombii* Torr.).

I. multiflora (Nutt.) V. Grant (*Gilia multiflora* Nutt.).

I. polyantha (Rydb.) V. Grant (*Gilia polyantha* Rydb.).

I. polycladon (Torr.) V. Grant (*Gilia polycladon* Torr.). Spreading Gilia.

I. pumila (Nutt.) V. Grant (*Gilia pumila* Nutt.). Low Gilia.

I. tenuituba (Rydb.) V. Grant (*Gilia tenuituba* Rydb.).

I. thurberi (Torr.) V. Grant (*Gilia thurberi* Torr.).

LANGLOISIA.

Langloisia matthewsii (Gray) Greene (?).

L. punctata (Coville) Goodding (?).

L. schottii (Torr.) Greene.

L. setosissima (T. & G.) Greene.

LEPTODACTYLON.

Leptodactylon pungens (Torr.) Nutt. ex Rydb.
    ssp. *brevifolium* (Rydb.) Wherry.

LINANTHUS.

Linanthus aureus (Nutt.) Greene.

L. aureus (Nutt.) Greene
    var. *decora* Gray.

L. bigelovii (Gray) Greene.

L. demissus (Gray) Greene.

L. dichotomus Benth. Evening Snow.

L. jonesii (Gray) Greene.

L. nuttallii (A. Gray) Greene ex Milliken (*Linanthastrum nuttallii* (Gray) Ewan) (157).

L. nuttallii (A. Gray) Greene ex Milliken
    ssp. *tenuilobus* R. Patterson (157).

LOESELIA.

Loeselia glandulosa (Cav.) G. Don.

MICROSTERIS.

Microsteris gracilis (Hook.) Greene.

NAVARRETIA.

Navarretia breweri (Gray) Greene.
N. minima Nutt.
N. propinqua Suksd.

PHLOX.

Phlox amabilis Brand.
P. austromontana Coville
    ssp. *densa* (Brand.) Wherry.
    ssp. *vera* Wherry.
P. diffusa Benth.
    ssp. *subcarinata* Wherry.
P. grayi Woot. & Standl.
P. griseola Wherry.
P. hoodii Richards.
P. hoodii Richards
    ssp. *canescens* (T. & G.) Wherry.
P. longifolia Nutt.
    ssp. *compacta* (Brand.) Wherry.
    ssp. *cortezana* (A. Nels.) Wherry.
    ssp. *longipes* (Jones) Wherry.
P. nana Nutt.
    ssp. *glabella* (Gray) Brand.
P. stansburyi (Torr.) Heller. Stansbury Phlox.
P. tenuifolia E. Nels.
P. woodhousei (Gray) E. Nels.

POLEMONIUM. Jacob's Ladder.

Polemonium delicatum Rydb.
P. filicinum Greene.
P. flavum Greene.
P. foliosissimum Gray.
P. foliosissimum Gray
    ssp. *robustum* (Rydb.) Brand.
P. pauciflorum Wats.
    ssp. *pauciflorum*
    ssp. *lemmonii* (Brand.) Wherry.

HYDROPHYLLACEAE. Water Leaf Family.

EMMENANTHE.

Emmenanthe penduliflora Benth. Whispering Bells.

ERIODICTYON. Yerba Santa.

Eriodictyon angustifolium Nutt. Narrow-leaved Yerba Santa.
E. angustifolium Nutt.
    var. *amplifolium* Brand.

EUCRYPTA.

Eucrypta chrysanthemifolia (Benth.) Greene
    var. *bipinnatifida* (Torr.) Constance. Torrey Eucrypta.
E. micrantha (Torr.) Heller. Small-flowered Eucrypta.

HESPEROCHIRON.

Hesperochiron pumilus (Dougl.) Porter.

HYDROPHYLLUM. Water Leaf.

Hydrophyllum occidentale (Wats.) Gray. Western Squaw Lettuce.

NAMA.

Nama demissum Gray.
N. demissum Gray
    var. *deserti* Brand. Purple Mat.
N. dichotomum (R. & P.) Choisy.
N. hispidum Gray
    var. *mentzelii* Brand. Sand Bells.
    var. *revolutum* Jeps.
    var. *spathulatum* (Torr.) C. L. Hitchc. Hispid Nama.
N. pusillum Lemmon. Small-leaved Nama.
N. retrorsum J. T. Howell.
N. rothrockii Gray.
N. stenocarpum Gray.

PHACELIA (158).

Phacelia affinis Gray. Purple Bell Phacelia.
P. alba Rydb. (*P. neomexicana* Thurber ex Torr. var. *alba*
    (*Rydb.*) Brand.).
P. ambigua Jones (*P. crenulata* Torr. var. ambigua (*Jones*)
    Macbr.). Notch-leaved Phacelia.
P. ambigua Jones
    var. *minutiflora* (Voss) Atwood.
P. arizonica Gray.

P. bombycina Woot. & Standl.

P. cephalotes Gray.

P. *coerulea* Greene.

*P. *constancei* Atwood.

P. corrugata A. Nels.

P. crenulata Torr.

P. crenulata Torr.
    var. *angustifolia* Atwood.

P. *cryptantha* Greene. Small-flowered Phacelia.

P. *cryptantha* Greene
    var. *derivata* Voss.

P. *demissa* Gray.

P. distans Benth.
    var. *australis* Brand. Wild Heliotrope.

P. filiformis Brand.

P. fremontii Torr. Fremont Phacelia.

P. glechomaefolia Gray.

P. integrifolia Torr. Crenate Leaf Phacelia.

P. ivesiana Torr. Ives Phacelia.

P. laxiflora J. T. Howell.

P. lemmoni Gray. Lemmon Phacelia.

P. magellanica (Lam.) Cov.

P. neglecta Jones.

P. neomexicana Thurber ex Torr.

P. palmeri Torr. ex Wats. (incl. var. *foetida* (Goodding) Brand.).

*P. *parryi* Torr. Parry Phacelia (96).

P. pedicellata Gray. Specter Phacelia.

P. pediculoides (J. T. Howell) Constance.

P. pulchella Gray.

P. *pulchella* Gray
    var. *goodingii* (Brand.) J. T. Howell.

*P. *rafaelensis* Atwood.

P. ramosissima Dougl.

P. rotundifolia Torr. Round-leaved Phacelia.

P. *rupestris* Greene (*P. congesta* Hook. var. *rupestris* (Greene)
    Macbr.). Blue Curls.

P. saxicola Gray.

P. *sericea* (Graham) A. Gray
    ssp. *ciliosa* (Rydb.) Gillett (159).

P. serrata Voss.

P. tanacetifolia Benth. Lacy Phacelia.

P. vallis-mortae Voss. Death Valley Phacelia.

*P. *welshii* Atwood.

PHOLISTOMA.

Pholistoma auritum (Lindl.) Lilja
  var. *arizonicum* (Jones) Constance.

TRICARDIA.

Tricardia watsoni Torr. Three Hearts.

BORAGINACEAE. Borage Family.

AMSINCKIA. Fiddle Neck.

Amsinckia intermedia F. & M. Coast Fiddleneck.
A. tessellata Gray. Checker Fiddleneck.

CORDIA.

Cordia parvifolia A. DC.

CRYPTANTHA (160).

Cryptantha abata Johnst.
C. albida (H.B.K.) Johnst.
C. angustifolia (Torr.) Greene. Narrow-leaved Cryptantha.
C. atwoodii Higgins (161).
C. bakeri (Greene) Payson.
C. barbigera (Gray) Greene. Bearded Cryptantha.
C barbigera (Gray) Greene
    var. *fergusonae* Macbr.
C. capitata (Eastw.) Johnst.
C. circumscissa (H. & A.) Johnst. Western Cryptantha.
C. confertiflora (Greene) Payson. Golden Cryptantha.
C. costata T. S. Brandeg. Ashen Cryptantha.
C. crassisepala (T. & G.) Greene. Thick-sepaled Cryptantha.
C. decipiens (Jones) Heller. Gravel Cryptantha.
C. dumetorum Greene. Flexuous Cryptantha.
C. fendleri (Gray) Greene.
C. flava (A. Nels.) Payson.
C. flavoculata (A. Nels.) Payson.
C. fulvocanescens (Wats.) Payson
      var. *fulvocanescens.*
      var. *echinoides* (Jones) Higgins.
C. gracilis Osterh. Slender Cryptantha.
C. holoptera (Gray) Macbr. Rough-stemmed Cryptantha.
C. humilis (A. Gray) Payson
    ssp. *ovina* (Payson) Higgins.
C. inaequata Johnst. Darwin Cryptantha.

C. jamesii (Torr.) Payson
 var. *disticha* (Eastw.) Payson.
 var. *multicaulis* (Torr.) Payson.
 var. *pustulosa* (Rydb.) Harringt.
 var. *setosa* (Jones) Johnst. ex Tidestr. (var. *cinerea*
 (Greene) Payson).
C. maritima Greene.
C. maritima Greene
 var. *pilosa* Johnst. White-haired Cryptantha.
C. micrantha (Torr.) Johnst. Purple-rooted Cryptantha.
C. muricata (H. & A.) Nels. & Macbr.
 var. *denticulata* (Greene) Jeps.
C. nevadensis Nels. & Kenn. Nevada Cryptantha.
C. nevadensis Nels. & Kenn.
 var. *rigida* Johnst.
C. osterhoutii (Payson) Payson (?).
C. pterocarya (Torr.) Greene. Wing Nut Cryptantha.
C. pterocarya (Torr.) Greene
 var. *cycloptera* (Greene) Macbr.
 var. *stenoloba* Johnst.
C. pusilla (T. & G.) Greene.
C. racemosa (Wats.) Greene. Woody Cryptantha.
C. recurvata Coville. Arched-calyxed Cryptantha.
C. semiglabra Barneby.
C. setosissima (Gray) Payson.
C. utahensis (Gray) Greene. Scented Cryptantha.
C. virginensis (Jones) Payson. Tufted Cryptantha.

HACKELIA. Stickseed.

Hackelia floribunda (Lehm.) Johnst.
H. pinetorum (Greene) Johnst.
H. ursina (Greene) Johnst.

HARPAGONELLA.

Harpagonella palmeri Gray
 var. *arizonica* Johnst.

HELIOTROPIUM. Heliotrope.

Heliotropium convolvulaceum (Nutt.) Gray. False Morning Glory.
H. convolvulaceum (Nutt.) Gray
 var. *californicum* (Greene) Johnst.
H. curassavicum L.
 var. *obovatum* DC. Alkali Heliotrope.
 var. *oculatum* (Heller) Johnst. Chinese Pusley.
H. phyllostachyum Torr.

LAPPULA. Stickseed.

Lappula echinata Gilib.
L. redowskii (Hornem.) Greene.
L. redowskii (Hornem.) Greene
    var. *desertorum* (Greene) Johnst.
L. texana (Scheele) Britt. Hairy Stickseed.
L. texana (Scheele) Britt.
    var. *coronata* (Greene) Nels. & Macbr.

LITHOSPERMUM. Gromwell, Puccoon.

Lithospermum cobrense Greene.
L. confine I. M. Johnst.
L. incisum Lehm.
L. multiflorum Torr.
L. viride Greene.

MACROMERIA.

Macromeria viridiflora DC.
M. viridiflora DC.
    var. *thurberi* (Gray) I. M. Johnst.

MERTENSIA. Bluebells.

Mertensia franciscana Heller.
M. macdougalii Heller.

MYOSOTIS. Scorpion Grass, Forget-Me-Not.

Myosotis scorpioides L. (25, 34).
M. verna Nutt. (13).

PECTOCARYA.

Pectocarya heterocarpa Johnst. Hairy-leaved Comb Bur.
P. platycarpa Munz & Johnst. Broad-nutted Comb Bur.
P. recurvata Johnst. Arch-nutted Comb Bur.
P. setosa Gray. Stiff-stemmed Comb Bur.

PLAGIOBOTHRYS.

Plagiobothrys arizonicus (Gray) Greene. Blood Weed, Arizona
    Popcorn Flower.
P. californicus Greene
    var. *fulvescens* Johnst.
P. cognatus (Greene) Johnst.
P. jonesii Gray. Jones Popcorn Flower.
P. pringlei Greene.
P. tenellus (Nutt.) Gray.

TIQUILIA (162, 163).

Tiquilia canescens (DC.) A. Richardson (*Coldenia canescens* DC.).
  Shrubby Coldenia.
  var. *canescens*
  var. *pulchella* (I. M. Johnst.) A. Richardson (*Coldenia
    canescens* DC var. *pulchella* I. M. Johnst.).
T. latior (I. M. Johnst.) A. Richardson (*Coldenia hispidissima*
    (T. & G.) Gray var. *latior* I. M. Johnst.).
T. nuttallii (Benth. ex Hook.) A. Richardson (*Coldenia nuttallii*
    Benth. ex Hook.). Nuttall Coldenia.
T. palmeri (A. Gray) A. Richardson (*Coldenia palmeri* Gray).
    Palmer Coldenia.
T. plicata (Torr.) A. Richardson (*Coldenia plicata* (Torr.)
    Coville). Plicate Coldenia.

VERBENACEAE. Vervain Family.

ALOYSIA.

Aloysia gratissima (Gill & Hook.) Troncoso
  var. *schulzae* (Standl.) Moldenke (*A. lycioides* Cham.
  var. *schulzae* (Standl.) Moldenke) (30).
A. wrightii (Gray) Heller. Oreganillo, Wright Lippa.

BOUCHEA.

Bouchea prismatica (L.) Kuntze.
B. prismatica (L.) Kuntze
  var. *brevirostra* Grenzebach.

LANTANA.

Lantana horrida H.B.K. Texas Lantana, Herbade Cristo.
L. macropoda Torr. Desert Lantana, Herba Negra, Mejorana.

PHYLA. Frog Fruit.

Phyla cuneifolia (Torr.) Greene. Wedge-leaf Frog Fruit.
P. incisa Small. Texas Frog Fruit.
P. lanceolata (Michx.) Greene. Northern Frog Fruit.
P. nodiflora (L.) Greene
  var. *rosea* (D. Don) Moldenke. Common Frog Fruit.

TETRACLEA.

Tetraclea coulteri Gray.
T. coulteri Gray
  var. *angustifolia* (Woot. & Standl.) Nels. & Macbr.

131

VERBENA. Vervain.

Verbena ambrosifolia Rydb. Western Pink Verbena.
V. ambrosifolia Rydb.
    forma *eglandulosa* Perry.
V. bipinnatifida Nutt. Dakota Vervain, Small-flowered
    Verbena.
V. bipinnatifida Nutt.
    var. *latilobata* Perry.
V. bracteata Lag. & Rodr. Prostrate Vervain.
V. carolina L.
V. ciliata Benth. Mexican Vervain, Alfrombrillo del Campo,
    Moradilla.
V. ciliata Benth.
    var. *pubera* (Greene) Perry.
V. ehrenbergiana Schauer (?).
V. gooddingii Briq. Goodding Verbena.
V. gooddingii Briq.
    var. *nepetifolia* Tidestrom.
V. gracilis Desf.
V. halei Small. Texas Vervain.
V. hastata L. Blue Vervain.
V. macdougalii Heller. New Mexican Vervain.
V. menthaefolia Benth. Mint Vervain.
V. neomexicana (Gray) Small. Hillside Vervain.
V. neomexicana (Gray) Small
    var. *xylopoda* Perry.
V. perennis Wooten. Pinleaf Vervain.
V. pinetorum Moldenke.
V. plicata Greene. Fanleaf Vervain.
V. pulchella Sweet
    var. *gracilior* (Troncoso) Shinners (V. tenuisecta Briq.) (164).
V. scabra Vahl. Sandpaper Vervain.
V. wrightii Gray. Desert Verbena.

LABIATAE. Mint Family.

AGASTACHE. Giant Hyssop.

Agastache barberi (Robins.) Epling.
A. breviflora (Gray) Epling.
A. micrantha (Gray) Woot. & Standl.
A. pallidiflora (Heller) Rydb.
    ssp *neomexicana* (Briq.) Lint & Epling.
    ssp. *typica* Lint & Epling.
A. rupestris (Greene) Standl.
A. wrightii (Greenm.) Woot. & Standl.

CLINOPODIUM. Wild Basil.

Clinopodium vulgare L.

HEDEOMA. Mock-Pennyroyal.

Hedeoma costatum Gray.
H. dentatum Torr.
H. diffusum Greene.
H. drummondii Benth.
H. hyssopifolium Gray.
H. nanum (Torr.) Briq.
    ssp. *californicum* Stewart.
    ssp. *macrocalyx* Stewart.
    ssp. *nanum*
H. oblongifolium (Gray) Heller.

HYPTIS. Desert Lavender.

Hyptis emoryi Torr.

LAMIUM. Dead Nettle.

Lamium amplexicaule L. Henbit.

LEONURUS. Motherwort.

Leonurus cardiaca L.

LYCOPUS. Bugleweed.

Lycopus americanus Muhl. Cutleaf Horehound.
L. lucidus Turez (?).

MARRUBIUM. Horehound.

Marrubium vulgare L. Common Horehound, Marrubio.

MENTHA. Mint.

Mentha arvensis L.
    var. *villosa* (Benth.) S. R. Stewart. Field Mint.
M. rotundifolia (L.) Hudson. Roundleaf Mint, Apple Mint.
M. spicata L. Spearmint.

MOLDAVICA. Dragon Head.

Moldavica parviflora (Nutt.) Britt.

MOLUCELLA. Molucca Balm.

Molucella laevis L. Shell Flower.

MONARDA. Bee Balm, Horse Mint.

Monarda austromontana Epling.
M. menthaefolia Graham.
M. pectinata Nutt. Prostrate Pink Horsemint.

MONARDELLA.

Monardella arizonica Epling.
M. odoratissima Benth.
    ssp. *parvifolia* (Greene) Epling.

NEPETA. Catnip.

Nepeta cataria L.

POLIOMINTHA. Rosemary Mint.

Poliomintha incana (Torr.) Gray.

PRUNELLA. Self Heal.

Prunella vulgaris L. Heal All.

SALAZARIA. Bladder Sage.

Salazaria mexicana Torr. Paper Bag Bush.

SALVIA. Sage.

Salvia aethiopis L.
S. amissa Epling.
S. arizonica Gray.
S. columbariae Benth. Chia.
S. davidsonii Greenm.
S. dorrii (Kell.) Abrams.
S. dorrii (Kell.) Abrams
    ssp. *argentea* (Rydb.) Munz.
    ssp. *mearnsii* (Britt.) McClintock.
S. henryi Gray.
S. lemmoni Gray.
S. mohavensis Greene. Mohave Sage.
S. pachyphylla Epl. Rose Sage.
S. parryi Gray.
S. pinguifolia (Fern.) Woot. & Standl. Rock Sage.
S. reflexa Hornem. Rocky Mountain Sage.
S. subincisa Benth.
S. tiliaefolia Vahl.

SCUTELLARIA. Skull Cap.

Scutellaria galericulata L. Marsh Skullcap.
S. lateriflora L. Mad Dog Skullcap.
S. potosina T. S. Brand.
  ssp.*platyphylla* Epling.
S. tessellata Epling.

STACHYS. Betony, Hedge Nettle.

Stachys agraria C. & S.
S. coccinea Jacq. Texas Betony.
S. palustris L.
  ssp. *pilosa* (Nutt.) Epling.
S. rothrockii Gray.

TEUCRIUM. Germander.

Teucrium canadense L. American Germander, Wood Sage.
T. canadense L.
  var. *angustatum* Gray.
  var. *occidentale* McClintock & Epling.
T. cubense Jacq.
  ssp. *depressum* (Small) McClintock & Epling. Small Coast
  Germander.
T. glandulosum Kell.

TRICHOSTEMA. Blue Curls.

Trichostema arizonicum Gray.
T. brachiatum L. Flux Weed.
T. micranthum Gray.

SOLANACEAE. Potato Family, Nightshade Family.

CAPSICUM. Red Pepper, Chillipiquin.

Capsicum baccatum L.

CHAMAESARACHA. False Nightshade.

Chamaesaracha coronopus (Dunal) Gray. Small Groundcherry.
C. sordida (Dunal) Gray (*C. coniodes* (Moric.) Britt.) (165).

DATURA. Thorn Apple, Jimson Weed.

Datura discolor Bernh. Desert Thorn Apple.
D. meteloides DC. Sacred Datura, Tolguacha.
D. quercifolia H.B.K. Oak Leaf Thorn Apple.
D. stramonium L. Jimson Weed, Toloache.

135

LYCIUM. Wolfberry, Desert Thorn.

Lycium andersonii Gray. Anderson Thornbush.

L. andersonii Gray
   var. *deserticola* C. L. Hitchc. Narrow-leaved Thornbush.
   var. *wrightii* Gray.

L. berlandieri Dunal
   var. *parviflorum* (Gray) Terrae.

L. californicum Nutt.

L. cooperi Gray. Peach Thorn.

L. exsertum Gray.

L. fremontii Gray. Fremont Thornbush.

L. macrodon Gray.

L. pallidum Miers. Rabbit Thorn.

L. parishii Gray. Parish Thornbush.

L. torreyi Gray. Squaw Thorn.

MARGARANTHUS.

Margaranthus lemmoni Gray.

M. solanaceus Schlecht. Netted Globe Berry.

NICOTIANA.

Nicotiana attenuata Torr.

N. clevelandi Gray.

N. glauca Graham. Tree Tobacco, Tronadora.

N. trigonophylla Dunal. Desert Tobacco, Tabaquillo.

N. trigonophylla Dunal
   var. *palmeri* (Gray) Jones (*N. palmeri* Gray) (166).

PETUNIA.

Petunia parviflora Juss. Wild Petunia.

PHYSALIS. Ground Cherry (168).

Physalis acutifolia (Miers) Sandw. (*P. wrightii* Gray.) Wright
   Ground Cherry (167).

P. angulata L.
   var. *lanceifolia* (Nees) Waterfall (*P. lanceifolia* Nees).

P. caudella Standl.

P. crassifolia Benth. (includes var. *cardiophylla* (Torr.) Gray).
   Thick-leaved Ground Cherry.

P. crassifolia Benth.
   var. *versicolor* (Rydb.) Waterfall (*P. versicolor* Rydb.).

P. foetens Poir.
   var. *neomexicana* (Rydb.) Waterfall (*P. neomexicana* Rydb.).

P. hederaefolia Gray. Ivy-leaved Ground Cherry.
P. hederaefolia Gray
    var. *cordifolia* (Gray) Waterfall (*P. fendleri* Gray). Fendler
      Ground Cherry.
P. latiphysa Waterfall.
P. lobata Torr. Purple Ground Cherry.
P. lobata Torr.
    forma *albiflora* Waterfall.
P. pubescens L.
    var. *integrifolia* (Dunal) Waterfall (*P. pubescens* L.).
    Hairy Ground Cherry, Tomate Fresadilla.
P. virginiana Miller
    var. *sonorae* (Torr.) Waterfall (*P. lanceolata* Michx.;
    P. longifolia Nutt.). Longleaf Ground Cherry.

SALPICHROA.

Salpichroa origanifolia (Lam.) Baill. *(S. rhomboidea* (Gill. & Hook.)
    Miers.). Lily-of-the-Valley Vine, Cocks Eggs (169).

SARACHA.

Saracha procumbens (Cav.) R. & P.

SOLANUM. Nightshade.

Solanum americanum Mill. American Nightshade, Hierba
    Mora Negra.
S. carolinense L. (50).
S. deflexum Greenm.
S. douglasii Dunal.
S. elaeagnifolium Cav. Silverleaf Nightshade, Trompillo.
S. fendleri Gray. Wild Potato.
S. heterodoxum Dunal
    var. *novomexicanum* Bartlett.
S. jamesii Torr. Wild Potato.
S. lumholtzianum Bartlett.
S. nodiflorum Jacq.
S. rostratum Dunal. Buffalo Bur, Mala Muier.
S. sarachoides Sendt.
S. sisymbriifolium Lam. Sticky Nightshade.
S. triflorum Nutt. Cutleaf Nightshade.
S. xanti Gray. Purple Nightshade.

SCROPHULARIACEAE.

ANTIRRHINUM. Snapdragon.

Antirrhinum cyathiferum Benth.
A. filipes Gray. Twining Snapdragon.
A. kingii Wats. King Snapdragon.
A. nuttallianum Benth.

BACOPA. Water Hyssop.

Bacopa eisenii (Kellogg) Pennell.
*B. monnieri (L.) Wettst. Water Hyssop (13).
B. rotundifolia (Michx.) Wettst. Disk Water Hyssop (25).

BESSEYA.

Besseya arizonica Pennell.
B. plantaginea (James) Rydb.

BRACHYSTIGMA.

Brachystigma wrightii (Gray) Pennell.

BUCHNERA. Blue Hearts.

Buchnera arizonica (Gray) Pennell.

CASTILLEJA. Indian Paint Brush.

Castilleja austromontana Standl. & Blumer.
C. chromosa A. Nels.
C. confusa Greene.
C. cruenta Standl.
C. exilis A. Nels.
C. integra Gray.
C. integra Gray
    var. *gloriosa* (Britt.) Ckll.
*C. kaibabensis N. Holmgren (170).
C. lanata Gray.
C. linariaefolia Benth. Long-leaved Paintbrush.
C. linariaefolia Benth.
    var. *omnipubescens* (Pennell) Clokey *forma omnipubescens*
    Pennell.
C. lineata Greene.
C. minor Gray.
C. mogollonica Pennell.
C. patriotica Fern.

C. patriotica Fern.
  var. *blumeri* (Standl.) K. & P.
C. sessiliflora Pursh. Downy Painted Cup.
*C. tenuiflora Benth. (171).

COLLINSIA.

Collinsia parviflora Dougl.

CORDYLANTHUS. Bird Beak, Club Flower.

Cordylanthus laxiflorus Gray.
C. nevinii Gray. Nevin Bird's Beak.
C. parviflorus (Ferris) Wiggins.
C. tenuifolius Pennell.
C. wrightii Gray.
C. wrightii Gray
  var. *pauciflorus* K. & P.

GRATIOLA. Hedge Hyssop.

Gratiola neglecta Torr.

KECKIELLA (172).

Keckiella antirrhinoides (Benth.) Straw
  ssp. *microphylla* (Gray) Straw (*Penstemon microphyllus* Gray).

LIMOSELLA. Mudwort.

*Limosella acaulis S. & M. (25, 34).
L. aquatica L.
L. pubiflora Pennell.

LINARIA. Toad Flax.

Linaria dalmatica Miller.
L. texana Scheele. Texas Toad Flax.
L. vulgaris Miller. Common Toad Flax, Butter and Eggs.

LINDERNIA.

Lindernia dubia (L.) Penn. (25) (?).

MAURANDYA.

Maurandya acerifolia Pennell.
M. antirrhiniflora H. & B. Blue Snapdragon Vine.
M. wislizeni Engelm.

MECARDONIA.

Mecardonia vandellioides (H.B.K.) Pennell.

MIMULUS. Monkey Flower.

Mimulus bigelovii Gray. Bigelow Mimulus.
M. cardinalis Dougl.
M. cardinalis Dougl.
    var. *verbenaceus* (Greene) K. & P.
M. dentilobus Robins. & Fern.
M. eastwoodiae Rydb.
M. floribundus Dougl.
M. glabratus H.B.K.
M. glabratus H. B. K.
    var. *fremontii* (Benth.) Grant.
M. guttatus DC.
M. guttatus DC.
    var. *gracilis* (Gray) Campbell.
M. nasutus Greene.
M. parryi Gray.
M. pilosus (Benth.) Wats.
M. primuloides Benth.
M. rubellus Gray. Red-stemmed Mimulus.
M. suksdorfii Gray.
M. tilingii Regel.

MOHAVEA.

Mohavea breviflora Coville. Lesser Mohavea.
M. confertiflora (Benth.) Heller. Ghost Flower.

ORTHOCARPUS. Owl Clover.

Orthocarpus luteus Nutt.
O. purpurascens Benth.
    var. *palmeri* Gray. Mohave Owl Clover.
O. purpureo-albus Gray.

PEDICULARIS. Lousewort, Wood Betony.

Pedicularis centranthera Gray.
P. grayi A. Nels.
P. groenlandica Retz.
    var. *surrecta* (Benth.) Piper. Elephant Head.
P. parryi Gray.
P. parryi Gray
    ssp. *mogollonica* (Greene) Carr (173).
P. racemosa Dougl.

PENSTEMON. Beard Tongue.

Penstemon albomarginatus Jones. White-margined Penstemon.
P. ambiguus Torr.
    ssp. *laevisissimus* Keck. Pink Plains Penstemon.
P. angustifolius Nutt.
    ssp. *venosus* Keck.
P. barbatus (Cav.) Roth.
P. barbatus (Cav.) Roth
    ssp. *torreyi* (Benth.) Keck.
    ssp. *trichander* (Gray) Keck.
*P. *barbatus x P. comarrhenus* (174).
*P. *barbatus x P. virgatus* (174).
P. bicolor (T. S. Brand.) Clokey & Keck
    ssp. *roseus* Clokey & Keck.
P. bridgesii Gray.
P. caespitosus Nutt.
    ssp. *desertipicti* (A. Nels.) Keck.
P. clutei A. Nels.
P. cobaea Nutt. Foxglove, Cobaea Penstemon.
P. commarrhenus Gray.
P. crideri A. Nels. (*P. pseudospectabilis* Jones *ssp. connatifolius*
    (A. Nels.) Keck *x P. eatonii* Gray ssp. *exertus* (A. Nels.) Keck).
P. dasyphyllus Gray.
P. discolor Keck.
P. eatoni Gray. Eaton Firecracker.
P. eatoni Gray
    ssp. *exsertus* (A. Nels.) Keck.
    ssp. *undosus* (Jones) Keck.
P. fendleri T. & G.
P. jamesii Benth.
    ssp. *ophianthus* (Pennell) Keck.
P. leritus Pennell.
P. linarioides Gray.
P. linarioides Gray
    ssp. *coloradensis* (A. Nels.) Keck.
    ssp. *compactifolius* Keck.
    ssp. *maguirei* Keck.
    ssp. *sileri* (Gray) Keck.
    ssp. *viridis* Keck.
P. minus A. Nels. (*P. eatoni* Gray ssp. *exertus* (A. Nels.) Keck *x P.
    palmeri* Gray).

141

PENSTEMON. Beard Tongue. *(cont.)*

P. nudiflorus Gray.
P. oliganthus Woot. & Standl.
P. pachyphyllus Gray
    ssp. *congestus* (Jones) Keck.
P. palmeri Gray. Scented Penstemon.
P. palmeri Gray
    ssp. *eglandulosus* Keck.
P. parryi Gray.
P. pinifolius Greene.
P. pseudospectabilis Jones. Mohave Beard Tongue.
P. pseudospectabilis Jones
    ssp. *connatifolius* (A. Nels.) Keck.
P. ramosus Crosswhite (incl. *P. lanceolatus* Benth. sensu
    K. & P.) (175).
P. rydbergii A. Nels.
P. stenophyllus Gray.
P. strictus Benth.
    ssp. *angustus* Pennell.
    ssp. *strictiformis* (Rydb.) Keck.
P. subulatus Jones.
P. superbus A. Nels.
P. thompsoniae (Gray) Rydb. Thompson Penstemon.
P. thurberi Torr. Thurber Penstemon.
P. utahensis Eastw. Utah Firecracker.
P. virgatus Gray.
P. virgatus Gray
    ssp. *arizonicus* (Gray) Keck.
P. watsoni Gray.
P. whippleanus Gray.

RHINANTHUS. Yellow Rattle.

Rhinanthus rigidus Chabert.

SCHISTOPHRAGMA.

Schistophragma intermedia (Gray) Pennell.

SCROPHULARIA.

Scrophularia californica Cham.
S. parviflora Woot. & Standl.

STEMODIA.

Stemodia durantifolia (L.) Swartz.

VERBASCUM. Mullein.

Verbascum blatteria L. (34).
V. *thapsus* L. Common Mullein.
V. *virgatum* Stokes.

VERONICA. Speedwell.

Veronica americana (Raf.) Schwein. American Brooklime.
V. *anagallis-aquatica* L. Water Speedwell.
V. *arvensis* L. Corn Speedwell.
V. *connata* Raf.
V. *peregrina* L.
    ssp. *xalapensis* (H.B.K.) Pennell. Necklace Weed.
V. *persica* Poir. Persian Speedwell.
V. *polita* Fries. Wayside Speedwell.
V. *serpyllifolia* L.
    var. *borealis* Laestad.
V. *wormskjoldii* R. & S.

BIGNONIACEAE. Bignonia Family.

CATALPA. Indian Bean.

Catalpa bignonioides Watt. Southern Catalpa (82).

CHILOPSIS. Desert Willow.

Chilopsis linearis (Cav.) Sweet
    var. *arcuata* Fosberg. Desert Catalpa, Mimbre.
    var. *glutinosa* (Engelm.) Fosberg.

TECOMA. Trumpet Flower.

Tecoma stans (L.) H.B.K.
    var. *angustata* Rehd. Yellow Trumpet Bush, Minona, Esperanza, Tronadora.

MARTYNIACEAE. Unicorn Plant Family.

PROBOSCIDEA. Devil's Claw, Cinco Llagas.

Proboscidea altheaefolia (Benth.) Decne. Desert Unicorn Plant, Elephant Tusks.
P. arenaria (Engelm.) Decne.
P. parviflora (Woot.) Woot. & Standl.

OROBANCHACEAE. Broom Rape Family.

CONOPHOLIS. Squaw Root.

Conopholis alpina Liebmann var. *mexicana* (Gray ex Watson) Haynes (C. *mexicana* Gray). Mexican Squaw Root (176).

OROBANCHE. Broom Rape.

Orobanche cooperi (Gray) Heller (*O. ludoviciana* Nutt. var. *cooperi* (Gray) G. Beck and var. *latiloba* Munz). Burro Weed Strangler.
O. fasciculata Nutt. Pinon Strangleroot.
O. fasciculata Nutt.
    var. *lutea* (Parry) Achey.
    var. *subulata* Goodman.
O. multicaulis Bdge. (50).
O. multiflora Nutt.
O. multiflora Nutt.
    var. *arenosa* (Suksdorf) Munz.
    var. *pringlei* Munz.

LENTIBULARIACEAE. Bladderwort Family.

UTRICULARIA. Bladderwort.

Utricularia vulgaris L. Common Bladderwort.

ACANTHACEAE. Acanthus Family.

ANISACANTHUS.

Anisacanthus thurberi (Torr.) Gray. Chuparosa, Desert Honeysuckle.

BELOPERONE.

Beloperone californica Benth. Chuparosa, Honeysuckle.

CARLOWRIGHTIA.

Carlowrightia arizonica Gray.
C. linearifolia (Torr.) Gray.

DICLIPTERA.

Dicliptera pseudoverticillaris Gray.
D. resupinata (Vahl) Juss.

DYSCHORISTE.

Dyschoriste decumbens (Gray) Kuntze.

ELYTRARIA. Scaly Stem.

Elytraria imbricata (Vahl) Pers. Purple Scaly Stem.

JACOBINIA.

Jacobinia ovata Gray.

RUELLIA.

Ruellia nudiflora (Engelm. & Gray) Urban.
R. nudiflora (Engelm. & Gray) Urban
     var. *glabrata* Leonard.

SIPHONOGLOSSA.

Siphonoglossa longiflora (Torr.) Gray.

TETRAMERIUM.

Tetramerium hispidum Nees.

PLANTAGINACEAE. Plantain Family.

PLANTAGO. Plantain, Indian Wheat.

Plantago argyraea Morris.
P. heterophylla Nutt. Many-seeded Plantain.
P. hirtella Kunth
     var. *mollior* Pilger.
P. insularis Eastw. Wooly Plantain.
P. lanceolata L. Buckhorn Plantain.
P. major L. Common Plantain.
P. purshii R. & S. Pursh Plantain.
P. purshii R. & S.
     var. *picta* (Morris) Pilger.
P. rhodosperma Decne. Red-seeded Plantain.
P. tweedyi Gray.
P. virginica L. Pale-seeded Plantain.
P. wrightiana Decne.

RUBIACEAE. Madder Family.

BOUVARDIA.

Bouvardia glaberrima Engelm.

CEPHALANTHUS. Button Bush.

Cephalanthus occidentalis L.
   var. *californicus* Benth. Common Button Bush.

CRUSEA.

Crusea subulata (Pavon) Gray.
C. wrightii Gray.

DIODIA. Button Weed.

Diodia teres Walt. Rough Buttonweed, Poor Joe.
D. teres Walt.
   var. *angustata* Gray.

GALIUM. Bedstraw, Cleavers.

Galium aparine L.
G. aparine L.
   var. vaillantii (DC.) Koch. Goosegrass.
G. bifolium Wats.
G. boreale L. Northern Bedstraw.
G. brandegei Gray.
G. collomae J. T. Howell.
G. coloradoense W. F. Wight.
G. fendleri Gray.
G. mexicanum H.B.K. (177).
      ssp. *asperrimum* (Gray) Demp.
      ssp. *mexicanum*
G. microphyllum Gray (*Relbunium microphyllum* (Gray) Hemsl.).
G. munzii Hilend & Howell. Munz Bedstraw.
G. munzii Hilend & Howell
      ssp. *ambivalens* Dempst. & Ehrend. (178).
G. obtusum Bigel. Blunt Leaf Bedstraw (25) (?).
G. pilosum Ait. Hairy Bedstraw.
G. proliferum Gray. Great Basin Bedstraw.
G. stellatum Kell.
      var. *eremicum* Hilend & Howell. Desert Bedstraw.
G. tinctorium L.
      var. *diversifolium* W. F. Wight.
      var. *subbiflorum* (Wieg.) Fern.
G. triflorum Michx. Fragrant Bedstraw.

G. triflorum Michx.
  forma glabrum Leyendecker.
G. watsoni (Gray) Heller.
G. wrightii Gray.
G. wrightii Gray
  var. *rothrockii* (Gray) Ehrendorfer.

HEDYOTIS (30).

Hedyotis greenei (Gray) W. H. Lewis (*Oldenlandia Greenei* Gray).
H. nigricans (Lam.) Fosb. (*Houstonia nigricans* (Lam.) Fern.).
H. pygmaea R. & S. (*Houstonia wrightii* Gray).
H. rubra (Cav.) Gray (*Houstonia rubra* Cav.).

KELLOGGIA.

Kelloggia galioides Torr.

MITRACARPUS.

Mitracarpus breviflorus Gray.

SHERARDIA (94).

°*Sherardia arvensis* L.

CAPRIFOLIACEAE. Honeysuckle Family.

LINNAEA. Twin Flower.

Linnaea borealis L.
  var. *americana* (Forbes) Rehd.

LONICERA. Honeysuckle.

Lonicera albiflora T. & G.
  var. *dumosa* (Gray) Rehd. White Honeysuckle.
L. arizonica Rehd. Arizona Honeysuckle.
L. interrupta Benth. Chaparral Honeysuckle.
L. involucrata (Richards.) Banks. Bearberry Honeysuckle.
L. japonica Thunb. Japanese Honeysuckle.
L. utahensis Wats.

SAMBUCUS. Elder, Elderberry.

Sambucus glauca Nutt. Blueberry Elder.
S. melanocarpa Gray. Blackbead Elder.
S. mexicana Presl. Tapiro Sauco, Mexican Elder.
S. microbotrys Rydb. Red Elderberry.
S. velutina Dur. & Hilg. Velvet Elder.

147

SYMPHORICARPOS. Snowberry.

Symphoricarpos longiflorus Gray. Long-flowered snowberry.
S. oreophilus Gray.
S. palmeri G. N. Jones.
S. parishii Rydb.
S. rotundifolius Gray.
S. utahensis Rydb.

VALERIANACEAE. Valerian Family.

PLECTRITIS.

Plectritis ciliosa (Greene) Jeps.
    ssp. *ciliosa*
    ssp. *insignis* (Suksd.) Morey.

VALERIANA. Valerian, Tobacco Root.

Valeriana arizonica Gray.
V. capitata Pall. ex Link.
    ssp. *acutiloba* (Rydb.) F. G. Meyer.
V. edulis Nutt.
V. occidentalis Heller.
V. sorbifolia H.B.K.

CUCURBITACEAE. Gourd Family.

APODANTHERA. Melon Loco.

Apodanthera undulata Gray.

BRANDEGEA.

Brandegea bigelovii (Wats.) Cogn. Brandegea.

CUCURBITA.

Cucurbita californica Torr.
C. digitata Gray. Finger-leaved Gourd.
°C. digitata Gray x *C. palmata* Wats. (179).
C. foetidissima H.B.K. Buffalo Gourd, Calabazilla.
C. palmata Wats. Coyote Melon.

CYCLANTHERA.

Cyclanthera dissecta (T. & G.) Arn. Bur Cucumber.

ECHINOCYSTIS. Mock Cucumber.

Echinocystis lobata (Michx.) T. & G. Wild Cucumber.

ECHINOPEPON. Wild Balsam Apple.

Echinopepon wrightii (Gray) Wats.

IBERVILLEA. Globe Berry.

Ibervillea tenuisecta (Gray) Small (?).

MARAH. Big Root, Wild Cucumber.

Marah gilensis Greene.

SICYOS. One-Seeded Bur Cucumber.

Sicyos ampelophyllus Woot. & Standl.
S. laciniatus L.
    var. *typica* Cogn.

SICYOSPERMA.

Sicyosperma gracile Gray.

TUMAMOCA.

Tumamoca macdougalii Rose.

CAMPANULACEAE. Bellflower Family.

CAMPANULA. Bell Flower.

Campanula parryi Gray.
C. rotundifolia L. Harebell, Bluebell.

LOBELIA.

Lobelia anatina Wimmer.
L. cardinalis L.
    ssp. *graminea* (Lam.) McVaugh. Cardinal Flower.
L. fenestralis Cav. Leafy Lobelia.
L. laxiflora H.B.K.
    var. *angustifolia* A. DC.

NEMACLADUS.

Nemacladus glanduliferus Jeps.
    var. *orientalis* McVaugh. Thread Plant.
N. gracilis Eastw. Small-flowered Thread Plant (?).
N. longiflorus Gray
    var. *breviflorus* McVaugh. Long-flowered Thread Plant.
N. rubescens Greene.
N. sigmoideus G. T. Robins. (50).

PORTERELLA.

Porterella carnosula (H. & A.) Torr.

TRIODANIS. Venus Looking Glass.

Triodanis biflora (R. & P.) Greene. Small Venus Looking Glass.
T. holzingeri McVaugh.
T. perfoliata (L.) Nieuwl. Venus Looking Glass.

COMPOSITAE. Sunflower Family.

ACAMPTOPAPPUS. Golden Head.

Acamptopappus sphaerocephalus (Harv. & Gray) Gray.

ACHAETOGERON.

Achaetogeron chihuahuensis Larsen.

ACHILLEA. Yarrow, Milfoil (180).

Achillea millefolium L.
    var. *alpicola* (Rydb.) Garrett (*A. lanulosa* Nutt. var.
    alpicola Rydb.).
    var. *lanulosa* (Nutt.) Piper (*A. lanulosa* Nutt.).

ACOURTIA (181).

Acourtia nana (Gray) Reveal & King (*Perezia nana* Gray).
    Desert Holly.
A. thurberi (Gray) Reveal & King (*Perezia thurberi* Gray).
A. wrightii (Gray) Reveal & King (*Perezia wrightii* Gray). Brownfoot.

AGERATINA.

Ageratina herbacea (Gray) King & Robins. (*Eupatorium herbaceum*
    (Gray) Greene) (182).

AGOSERIS. Mountain Dandelion.

Agoseris arizonica Greene.
A. aurantiaca Greene.
A. glauca (Pursh) D. Dietr.
    var. *dasycephala* (T. & G.) Jeps.
    var. *laciniata* (D. C. Eaton) Smiley.
    var. *parviflora* (Nutt.) Rydb.
A. heterophylla (Nutt.) Greene.

AMBROSIA. Ragweed, Bur Sage (183).

Ambrosia acanthicarpa Hook. (*Franseria acanthicarpa* (Hook.)
    Coville). Annual Burweed.
A. ambrosioides (Cav.) Payne (*Franseria ambrosioides* Cav.).

A. aptera DC. Blood Weed.
A. artemisiifolia L. Ragweed, Altamisa.
A. confertiflora DC. (*Franseria confertiflora* (DC.) Rydb.).
   Slimleaf Bursage.
A. cordifolia (Gray) Payne (*Franseria cordifolia* Gray).
A. deltoidea (Torr.) Payne (*Franseria deltoidea* Torr.). Burrobush.
A. dumosa (Gray) Payne (*Franseria dumosa* Gray). White Bursage.
A. eriocentra (Gray) Payne (*Franseria eriocentra* Gray). Woolly-
   fruited Bursage.
A. ilicifolia (Gray) Payne (*Franseria ilicifolia* Gray).
   Holly-leaved Bursage.
A. psilostachya DC. Western Ragweed.
A. tomentosa Nutt. (*Franseria discolor* Nutt.). Skeleton Leaf
   Bursage.

AMPHIPAPPUS. Chaff Bush.

Amphipappus fremontii T. & G.
   var. *spinosus* (A. Nels.) C. L. Porter. Eytelia.

ANAPHALIS. Pearly Everlasting.

Anaphalis margaritacea (L.) Gray.

ANISOCOMA. Scale Bud.

Anisocoma acaulis T. & G.

ANTENNARIA. Pussy Toes.

Antennaria arida E. Nels.
A. marginata Greene.
A. parvifolia Nutt.
A. rosulata Rydb.
A. umbrinella Rydb.

ANTHEMIS. Camomile.

Anthemis cotula L. Mayweed, Dog Fennel.

APHANOSTEPHUS. White Daisy.

Aphanostephus humilis (Benth.) Gray. Poorland Daisy.

ARCTIUM. Burdock.

Arctium minus Schk.

ARNICA.

Arnica chamissonis Less.
   ssp. *foliosa* (Nutt.) Maguire (*Arnica foliosa* Nutt.) (184).
A. cordifolia Hook.

ARTEMISIA. Wormwood, Sagebrush.

Artemisia annua L.
A. *arbuscula* Nutt.
    ssp. *nova* (A. Nels.) G. H. Ward.
A. *biennis* Willd.
A. *bigelovii* Gray.
A. *carruthii* Wood.
A. *carruthii* Wood
    var. *wrightii* (Gray) Blake.
A. *dracunculus* L. (*A. dracunculoides* Pursh and var. *dracunulina* (Wats.) Blake). (185).
A. *filifolia* Torr. Sand Sagebrush.
A. *franserioides* Greene.
A. *frigida* Willd. Estafiata, Prairie Sagewort.
A. *ludoviciana* Nutt. (incl. ssp. *albula* (Wooton) Keck; ssp. *mexicana* (Willd.) Keck; ssp. *redolens* (Gray) Keck; ssp. *sulcata* (Rydb.) Keck) (30).
A. *pacifica* Nutt.
A. *pygmaea* Gray. Pigmy Sagebrush.
A. *spinescens* D. C. Eaton. Bud Sagebrush.
A. *tridentata* Nutt. Big Sagebrush.

ASTER.

Aster adscendens Lindl.
A. *coerulescens* DC.
A. *commutatus* (T. & G.) Gray. White Prairie Daisy.
A. *commutatus* (T. & G.) Gray
    var. *crassulus* (Rydb.) Blake.
    var. *polycephalus* (Rydb.) Blake.
A. *foliaceus* Lindl.
    var. *burkei* Gray.
A. *frondosus* (Nutt.) T. & G.
A. *glaucodes* Blake.
A. *glaucodes* Blake
    var. *pulcher* Blake.
A. *intricatus* (Gray) Blake. Shrubby Alkali Aster.
A. *lemmoni* Gray.
A. *oregonus* (Nutt.) T. & G.
A. *pauciflorus* Nutt. Marsh Alkali Aster.
A. *riparius* H.B.K.
A. *spinosus* Benth. Spiny Aster, Mexican Devil Weed.
A. *subulatus* Michx.
    var. *ligulatus* Shinners (A. *exilis* Ell.) (30).

ATRICHOSERIS. Tobacco Weed.

Atrichoseris platyphylla Gray. Parachute Plant.

BACCHARIS. Groundsel Tree.

Baccharis bigelovii Gray.
B. brachyphylla Gray. Short-leaved Baccharis.
B. emoryi Gray. Emory Baccharis.
B. neglecta Britt. Linearleaf Baccharis, Jara Dulce.
B. pteronioides DC. Yerba-de-Pasmo.
B. salicifolia (R. & P.) Pers. (*B. glutinosa* Pers.). Seep Willow,
    Batamote, Jara (186).
B. sarathroides Gray. Broom Baccharis, Desert Broom.
B. sergiloides Gray. Waterweed.
B. thesioides H.B.K.
B. viminea DC. Mule Fat.
B. wrightii Gray. Wright's Baccharis.

BAHIA (187).

Bahia absinthifolia Benth.
    var. *dealbata* Gray and intermediates to var. *absinthifolia*.
B. biternata Gray.
B. dissecta (Gray) Britt. Yellow Ragweed.
B. oppositifolia (Nutt.) DC. (?).
B. woodhousei Gray.

BAILEYA. Desert Marigold.

Baileya multiradiata Harv. & Gray. Wild Marigold, Desert Baileya.
B. pauciradiata Harv. & Gray. Lax Flowers.
B. pleniradiata Harv. & Gray. Woolly Marigold.

BEBBIA. Sweet Bush.

Bebbia juncea (Benth.) Greene. Chuckawalla's Delight.
B. juncea (Benth.) Greene
    var. *aspera* Greene.

BERLANDIERA. Green Eyes.

Berlandiera lyrata Benth. Lyreleaf.
B. lyrata Benth.
    var. *macrophylla* Gray.

BIDENS. Bur Marigold, Beggar Tricks.

Bidens aurea (Ait.) Sherff
B. aurea (Ait.) Sherff
    var. *wrightii* (Gray) Sherff.
B. bigelovii Gray.
B. bipinnata L. Spanish Needles.
*B. cernua L. (148).
B. ferulaefolia (Jacq.) DC.
B. frondosa L. Sticktight, Beggar Ticks.
B. heterosperma Gray.
B. laevis (L.) B.S.P. Bur Marigold.
B. lemmoni Gray.
B. leptocephala Sherff.
B. pilosa L.
B. pilosa L.
    var. *radiata* Schultz Bip.
B. tenuisecta Gray.

BRICKELLIA.

Brickellia amplexicaulis Robins.
B. amplexicaulis Robins.
    var. *lanceolata* (Gray) Robins.
B. atractyloides Gray.
B. baccharidea Gray.
B. betonicaefolia Gray.
B. brachyphylla Gray.
B. californica (T. & G.) Gray. Pachaba.
B. chlorolepis (Woot. & Standl.) Shinners (*Kuhnia rosmarinifolia*
    Vent. and var. *chlorolepis* (Woot. & Standl.) Blake) (188).
B. coulteri Gray.
B. desertorum Coville. Desert Brickellia.
B. fendleri Gray.
B. floribunda Gray.
B. frutescens Gray. Shrubby Brickellia.
B. grandiflora (Hook.) Nutt. Large-flowered Thoroughwort.
B. incana Gray. Woolly Brickellia.
B. lemmoni Gray.
B. longifolia Wats.
B. multiflora Kellogg. Gum-leaved Brickellia.
B. oblongifolia Nutt.
    var. *linifolia* (D. C. Eaton) Robins. Pinon Brickellia.
B. parvula Gray.
B. pringlei Gray.
B. rusbyi Gray.

B. scabra (Gray) A. Nels.
B. simplex Gray.
B. squamulosa Gray.
B. venosa (Woot. & Standl.) Robins.

CALYCOSERIS.

Calycoseris parryi Gray. Yellow Jack Stem.
C. wrightti Gray. White Tack Stem.

CARDUUS. Plumeless Thistle.

Carduus nutans L. Musk Thistle.

CARMINATIA. Plume Weed.

Carminatia tenuiflora DC.

CARPHOCHAETE. Bristlehead.

Carphochaete bigelovii Gray.

CENTAUREA. Starthistle, Knapweed.

Centaurea americana Nutt. Sultan Starthistle, Cardo del Valle.
C. calcitrapa L.
C. cyanus L. Bachelor's Button.
°*C. maculosa* Lam. (56).
C. melitensis L. Malta Star Thistle, Tocolote.
C. repens L. Russian Knapweed.
C. rothrockii Greenm.
C. salmantica L.
C. solstitialis L. Yellow Star Thistle.

CHAENACTIS.

Chaenactis carphoclinia Gray. Pebble Pincushion.
C. carphoclinia Gray
   var. *attenuata* (Gray) Jones.
C. douglasii (Hook.) H. & A. Douglas Pincushion.
C. fremontii Gray. Fremont Pincushion.
C. macrantha D. C. Eaton.
C. stevioides H. & A. Esteve Pincushion.
C. stevioides H. & A.
   var. *brachypappa* (Gray) H. M. Hall.
   var. *thornberi* Stockwell.
C. xantiana Gray. Xantus Pincushion.

CHAMAECHAENACTIS.

Chamaechaenactis scaposa (Eastw.) Rydb.

155

CHAPTALIA.

Chaptalia alsophila Greene.

C. leucocephala Greene.

CHRYSANTHEMUM.

Chrysanthemum coronarium L. Crown Daisy.

C. leucanthemum L.
    var. *pinnatifidum* Lecoq. & Lam. Common Ox Eye Daisy.

CHRYSOPSIS (HETEROTHECA). Golden Aster.

Chrysopsis rutteri (Rothr.) Greene.

Note: No combination was presented for this taxon in the merger of
    Heterotheca and Chrysopsis (208).

CHRYSOTHAMNUS. Rabbit Brush, Rayless Goldenrod (189, 190).

Chrysothamnus depressus Nutt. Rabbit Brush.

C. greenei (Gray) Greene.

C. molesta (Blake) L. C. Anderson (*C. viscidiflorus* (Hook.)
    Nutt. var. *molestus* Blake).

C. nauseosus (Pall.) Britt.
    ssp. *bigelovii* (Gray) Hall & Clem.
    ssp. *consimilis* (Greene) Hall & Clem.
    ssp. *gnaphalodes* (Greene) Hall & Clem.
    ssp. *graveolens* (Nutt.) Piper.
    ssp. *junceus* (Greene) Hall & Clem.
    ssp. *latisquameus* (Gray) Hall & Clem.
    ssp. *leiospermus* (Gray) Hall & Clem.

C. paniculatus (Gray) H. M. Hall. Black-banded Rabbit Brush.

C. parryi (Gray) Greene
    ssp. *nevadensis* (Gray) Hall & Clem.

C. pulchellus (Gray) Greene.

C. teretifolius (Dur. & Hilg.) H. M. Hall.

C. viscidiflorus (Hook.) Nutt. Sticky-leaved Rabbit Brush.

C. viscidiflorus (Hook.) Nutt.
    ssp. *elegans* (Greene) Hall & Clem.
    ssp. *planifolius* L. C. Anderson.
    ssp. *stenophyllus* (Gray) Hall & Clem.

CICHORIUM. Chicory.

Cichorium intybus L. Blue Sailors.

CIRSIUM. Thistle.

Cirsium arizonicum (Gray) Petrak. Arizona Thistle.

C. arvense L.

  var. *mite* Wimm. & Grab. Canada Thistle.

C. bipinnatum (Eastw.) Rydb.

C. coloradense (Rydb.) Ckll. This is the first published report
of this addition to the flora of Arizona.

C. foliosum (Hook.) DC.

C. grahami Gray.

C. neomexicanum Gray.

C. nidulum (Jones) Petrak.

C. ochrocentrum Gray. Yellow Spine Thistle.

C. parryi (Gray) Petrak.

C. pulchellum (Greene) Woot. & Standl.

C. rothrockii (Gray) Petrak.

C. rydbergii Petrak.

C. undulatum (Nutt.) Spreng. Wavyleaf Thistle.

C. vulgare (Savi) Airy-Shaw. Bull Thistle.

C. wheeleri (Gray) Petrak.

C. wrightii Gray.

CNICUS. Blessed Thistle.

Cnicus benedictus L.

CONYZA (191).

Conyza bonariensis (L.) Cronq. (*Erigeron linifolius* Willd.).

C. canadensis (L.) Cronq. (*E. canadensis* L.). Horseweed.

C. coulteri Gray.

C. schiedeana (Less.) Cronq. (*Erigeron schiedeanus* Less.).

C. sophiaefolia H.B.K.

COREOCARPUS.

Coreocarpus arizonicus (Gray) Blake.

COREOPSIS. Tickseed.

Coreopsis atkinsoniana Dougl.

C. californica (Nutt.) Sharsmith (*C. douglasii* (DC.) Pall.) (192).

C. tinctoria Nutt. (*C. cardaminefolia* (DC.) T. & G.). Manzanilla
Silvestre, Calliopsis, Golden Coreopsis (193).

COSMOS.

Cosmos parviflorus (Jacq.) Pers.

COTULA.

Cotula australis (Sieber) Hook. f.
C. coronopifolia L. Brass Buttons.

CREPIS. Hawk Beard.

Crepis acuminata Nutt.
C. glauca (Nutt.) T. & G.
    ssp. *barberi* (Greenm.) Babcock & Stebbins.
    ssp. *glauca* Babcock & Stebbins.
C. intermedia Gray.
C. occidentalis Nutt. Western Crepis.
C. occidentalis Nutt.
    ssp. *costata* (Gray) Babcock & Stebbins.

CYNARA. Globe Artichoke.

Cynara scolymus L.

DICORIA.

Dicoria brandegei Gray. Single-fruited Dicoria.
D. canescens Gray. Desert Dicoria.

DIMORPHOTHECA.

Dimorphotheca aurantiaca DC. African Daisy (194).

DUGALDIA.

Dugaldia hoopesii (Gray) Rydb. (*Helenium hoopesii* Gray).
    Western Sneezeweed (195).

DYSSODIA. Dogweed, Fetid Marigold (30, 196).

Dyssodia acerosa DC.
D. concinna (Gray) Robins.
D. cooperi Gray. Cooper Dyssodia.
D. neomexicana (Gray) Robins.
D. papposa (Vent.) Hitchc. False Dogfennel.
D. pentachaeta (DC.) Robins.
    var. *belenidium* (DC.) Strother (*D. thurberi* (Gray) A. Nels.).
    Thurber Dyssodia.
    var. *hartwegii* (Gray) Strother (*D. hartwegii* (Gray) Robins.).
D. porophylloides Gray. San Felipe Dyssodia.

ECLIPTA.

Eclipta alba (L.) Hassk. Yerba de Tajo.

ENCELIA.

Encelia farinosa Gray. Brittle Bush, Incienso.
E. farinosa Gray
    var. *phenicodonta* (Blake) Johnst.
°*E. farinosa* Gray x *Geraea canescens* T. & G. (197).
E. frutescens Gray. Rayless Encelia.
E. frutescens Gray
    var. *resinosus* Jones.
E. virginensis A. Nels. (*E. frutescens* Gray var. *virginensis* (A. Nels.)
    Blake) (198).
E. virginensis A. Nels.
    ssp. *actoni* (Elmer) Keck (*E. frutescens* Gray var. *actoni* (Elmer)
    Blake) (198).

ENCELIOPSIS.

Enceliopsis argophylla (D.C. Eaton) A. Nels.
E. nudicaulis (Gray) A. Nels. Naked-stemmed Sunray.

ENGELMANNIA.

Engelmannia pinnatifida Nutt. Engelmann's Daisy.

ERICAMERIA.

Ericameria cuneata (Gray) McClatchie
    var. *spathulata* (Gray) Hall (*Haplopappus cuneatus* Gray var.
    spathulata (Gray) Blake). Desert Rock Goldenbush (199).
E. laricifolia (Gray) Shinners (*Haplopappus laricifolius* Gray).

ERIGERON. Fleabane.

Erigeron arizonicus Gray.
E. bellidiastrum Nutt. Western Fleabane.
E. caespitosus Nutt.
E. canus Gray.
E. compactus Blake
    var. *consimilis* (Cronq.) Blake.
E. concinnus (H. & A.) T. & G. Tidy Fleabane.
E. concinnus (H. & A.) T. & G.
    var. *aphanactis* Gray.
    var. *condensatus* D. C. Eaton.
E. divergens T. & G. Spreading Fleabane.
E. eatoni Gray.
E. eriophyllus Gray.
E. flagellaris Gray. Running Fleabane.

159

ERIGERON. Fleabane. *(cont.)*

E. *formosissimus* Greene
    var. *formosissimus.*
    var. *viscidus* (Rydb). Cronq.
E. *kuschei* Eastw.
E. *lemmoni* Gray.
E. *lobatus* A. Nels.
E. *lonchophyllus* Hook. (149).
E. *macranthus* Nutt.
E. *modestus* A. Gray (E. *nudiflorus* Buckl.) (30).
E. *neomexicanus* Gray.
E. *oreophilus* Greenm.
E. *oxyphyllus* Greene.
E. *perglaber* Blake (?).
E. *platyphyllus* Greene.
E. *pringlei* Greene.
E. *pusillus* Nutt.
E. *rusbyi* Gray.
E. *simplex* Greene.
E. *superbus* Greene.
E. *ursinus* D. C. Eaton.
E. *utahensis* Gray.
E. *utahensis* Gray
    var. *sparsifolius* (Eastw.) Cronq.
    var. *tetrapleurius* (Gray) Cronq.

ERIOPHYLLUM.

Eriophyllum confertiflorum (DC.) Gray. Golden Yarrow.
E. *lanosum* Gray. Woolly Eriophyllum.
E. *multicaule* (DC.) Gray.
E. *pringlei* Gray. Pringle Eriophyllum.
E. *wallacei* Gray. Wallace Eriophyllum.

EUPATORIUM. Thoroughwort, Boneset.

Eupatorium bigelovii Gray.
E. *greggii* Gray.
E. *lemmoni* Robins.
E. *maculatum* L. Joe Pye Weed.
E. *pauperculum* Gray.
E. *pycnocephalum* Less.
E. *rothrockii* Gray.
E. *solidaginifolium* Gray.
E. *wrightii* Gray.

EURYOPS.

Euryops multifidus (L. f.) DC.

EVAX. Rabbit Tobacco, Cotton Rose.

Evax multicaulis DC.

FILAGO.

Filago arizonica Gray. Arizona Filago.
F. californica Nutt.
F. depressa Gray. Dwarf Filago.

FLAVERIA.

* *Flaveria mcdougallii* Theroux, Pinkava & Keil (200).
F. trinervia (Spreng.) C. Mohr.

FLOURENSIA. Tar Bush, Varnish Bush.

Flourensia cernua DC. Black Brush, Hojase.

GAILLARDIA. Blanket Flower.

Gaillardia arizonica Gray.
G. arizonica Gray
    var. *pringlei* (Rydb.) Blake.
**G. multiceps* Greene (30).
G. parryi Greene.
G. pinnatifida Torr.
G. pulchella Foug. Indian Blanket, Firewheel.

GALINSOGA. Quick Weed.

Galinsoga parviflora Cav. (*Galinsoga semicalva* (Gray) St. John &
    White incl. var. *percalva* Blake) (201).

GERAEA. Desert Sunflower.

Geraea canescens T. & G. Hairy-headed Sunflower.
G. canescens T. & G.
    var. *paniculata* (Gray) Blake.

GLYPTOPLEURA.

Glyptopleura setulosa Gray. Keysia.

GNAPHALIUM. Cudweed.

Gnaphalium arizonicum Gray. Arizona Cudweed.
G. chilense Spreng. Small-flowered Cudweed, Cotton Batting.
G. exilifolium A. Nels. (*G. grayi* Nels. & Macbr.) (202).
G. leucocephalum Gray. White Cudweed.

GNAPHALIUM. Cudweed. *(cont.)*

G. luteo-album L.

G. macounii Greene. Clammy Cudweed.

G. palustre Nutt. Lowland Cudweed.

G. pringlei Gray.

G. purpureum L. Purple Cudweed.

G. wrightii Gray.

GREENELLA.

Greenella arizonica Gray.

G. discoidea Gray.

GRINDELIA. Gum Weed.

Grindelia aphanactis Rydb.

G. arizonica Gray.

G. arizonica Gray
  var. *microphylla* Steyerm.

G. gymnospermoides (A. Gray) Ruffin (*Xanthocephalum gymnospermoides* (Gray) Benth. & Hook.) (203).

G. laciniata Rydb.

G. squarrosa (Pursh) Dunal. Curly Cup Gumweed.

G. squarrosa (Pursh) Dunal
  var. *serrulata* (Rydb.) Steyerm.

GUARDIOLA.

Guardiola platyphylla Gray.

GUTIERREZIA. Snake Weed, Broom Weed (204).

Gutierrezia microcephala (DC.) Gray (incl. *G. linoides* Greene & *G. lucida* Greene). Three Leaf Snakeweed.

G. sarothrae (Pursh) Britt. & Rusby. Broom Snakeweed.

G. serotina Greene (*G. californica* (DC.) T. & G. sensu K. & P.).

GYMNOSPERMA.

Gymnosperma glutinosum (Spreng.) Less. (*Selloa glutinosa* Spreng.). Tatalencho (205).

HAPLOPAPPUS.

Note: For those who do not recognize Haplopappus as a component of the North American flora, i.e. only in South America, current names in other genera (where available) are given here and elsewhere in the list of genera.

Haplopappus acaulis (Nutt.) Gray
  var. *glabratus* DC. (?)=*Stenotus*.

H. acradenius (Greene) Blake=*Isocoma*. Alkali Goldenbush.

162

H. armerioides (Nutt.) Gray=*Stenotus*.

H.*cervinus* Wats. (no other current name is available).

°H. ciliatus (Nutt.) DC. (56)=*Prionopsis*.

H. croceus Gray=*Pyrrocoma*.

var.*genuflexus* (Greene) Blake (transfer not published).

H. drummondii (T. & G.) Blake=*Isocoma*.

H. heterophyllus (Gray) Blake=*Isocoma*. Jimmy Weed,
Rayless Goldenrod.

H. junceus Greene (?)=*Machaeranthera*.

H. laricifolius Gray=*Ericameria*. Turpentine Bush.

H. linearifolius DC.

var. *interior* (Coville) M.E. Jones=*Stenotopsis*.

H. nuttallii T. & G.=*Machaeranthera*.

H. parryi Gray=*Solidago*.

H. ravenii Jackson (34) (incl. in *Machaeranthera gracilis* (Nutt.)
Gray).

H. salicinus Blake (the affinities of this are not known).

H. scopulorum (Jones) Blake (no other current name is available).

H. scopulorum (Jones) Blake
var. *hirtellus* Blake.

H. spinulosus (Pursh) DC.=*Machaeranthera*.

var. *gooddingii* (A. Nels.) Blake=*Machaeranthera*.

var. *turbinellus* (Rydb.) Blake=*Machaeranthera*.

H. suffruticosus (Nutt.) Gray (?) (no other current name is available).

H. tenuisectus (Greene) Blake=*Isocoma*. Burroweed.

H. watsonii Gray (no other current name is available).

HELENIUM. Sneeze Weed.

Helenium arizonicum Blake.

H. autumnale L.

H. autumnale L.
var. *canaliculatum* (Lam.) T. & G.

H. thurberi Gray.

HELIANTHELLA.

Helianthella microcephala Gray.

H. parryi Gray.

H. quinquenervis (Hook). Gray.

HELIANTHUS. Sunflower (206).

Helianthus annuus L. Common Sunflower, Mirasol.

H. anomalus Blake.

H. arizonensis R. Jackson.

H. ciliaris DC. Plains Sunflower.

H. deserticola Heiser.

HELIANTHUS. Sunflower (205) *(cont.)*

H. niveus (Benth.) Brandegee
    ssp. *canescens* (A. Gray) Heiser (*H. petiolaris* Nutt
    var. *canescens* Gray).
    ssp. *tephrodes* (A. Gray) Heiser.
H. nuttallii T. & G.
H. petiolaris Nutt.
    ssp. *fallax* Heiser.

HELIOPSIS. Ox Eye.

Heliopsis parvifolia Gray.

HEMIZONIA. Tarweed.

Hemizonia kelloggii Greene. Kellogg Tarweed.
H. pungens (H. & A.) T. & G. Spikeweed.

HETEROSPERMA.

Heterosperma pinnatum Cav.

HETEROTHECA (CHRYSOPSIS). Telegraph Plant (207, 208).

Heterotheca fulcrata (Greene) Shinners (*Chrysopsis fulcrata*
    Greene).
H. grandiflora Nutt.
H. latifolia Buckley
    var. *latifolia*
    var. *macgregoris* Wagenkn.
H. psammophila Wagenkn. (*H. subaxillaris* (Lam.) Britt & Rusby
    sensu K. & P.) Camphor Weed.
H. villosa (Pursh) Shinners (*Chrysopsis villosa* (Pursh) Nutt.).
    Hairy Golden Aster.
H. villosa (Pursh) Shinners
    var. *foliosa* (Nutt.) Harms (*Chrysopsis foliosa* Nutt.).
    var. *hispida* (Hook.) Harms (*Chrysopsis hispida* (Hook.) DC.).
    Hispid Golden Aster.
H. viscida (A. Gray) Harms (*Chrysopsis viscida* (A. Gray) Greene).

HIERACIUM. Hawkweed.

Hieracium carneum Greene.
H. fendleri Schultz Bip.
H. fendleri Schultz Bip.
    var. *discolor* Gray.
    var. *mogollense* Gray.
H. lemmoni Gray.
H. pringlei Gray.
H. rusbyi Greene.

HYMENOCLEA. Burro Brush.

Hymenoclea monogyra T. & G.
H. salsola T. &. G. Cheesebush.
H. salsola T. & G.
    var. *patula* (A. Nels.) Peterson & Payne (209).
    var. *pentalepis* (Rydb.) Benson (210).

HYMENOPAPPUS.

Hymenopappus filifolius Hook.
    var. *cinereus* (Rydb.) Johnst. Yellow Cutleaf.
    var. *lugens* (Greene) Jeps.
    var. *megacephalus* B. L. Turner.
    var. *nanus* (Rydb.) B. L. Turner.
    var. *pauciflorus* (Johnst.) B. L. Turner.
H. flavescens Gray
    var. *cano-tomentosus* Gray.
H. mexicanus Gray.
H. radiatus Rose.

HYMENOTHRIX.

Hymenothrix loomisii Blake.
H. wislizenii Gray.
H. wrightii Gray.

HYMENOXYS. Bitterweed, Actinea.

Hymenoxys acaulis (Pursh) K. F. Parker.
    var. *arizonica* (Greene) K. F. Parker. Arizona Actinea.
H. argentea (Gray) K. F. Parker.
H. bigelovii (Gray) K. F. Parker.
H. brandegei (Porter) K. F. Parker.
H. cooperi (Gray) Ckll.
H. cooperi (Gray) Ckll.
    var. *canescens* (D. C. Eaton) K. F. Parker.
H. helenioides (Rydb.) Ckll.
H. ivesiana (Greene) K. F. Parker.
H. odorata DC. Bitterweed.
H. quinquesquamata Rydb.
H. richardsoni (Hook.) Ckll.
    var. *floribunda* (Gray) K. F. Parker. Pinque.
H. rusbyi (Gray) Ckll.
H. subintegra Ckll.

ISOCOMA.

Isocoma acradenia (Greene) Greene (*Haplopappus acradenius* (Greene) Blake).

I. drummondii (T. & G.) Greene (*Haplopappus drummondii* (T. & G.) Blake).

I. tenuisecta Greene (*Haplopappus tenuisectus* (Greene) Blake).

I. wrightii (Gray) Rybd. (*Haplopappus heterophyllus* (Gray) Blake).

IVA. Marsh Elder.

Iva acerosa (Nutt.) Jackson (*Oxytenia acerosa* Nutt.). Copperweed (211).

I. ambrosiaefolia Gray.

I. axillaris Pursh. Poverty Weed.

I. dealbata Gray. Woolly Sump Weed (55).

I. xanthifolia Nutt. Burweed Marsh Elder.

KRIGIA.

Krigia biflora (Walt.) Blake.

LACTUCA. Lettuce.

Lactuca graminifolia Michx.

L. ludoviciana (Nutt.) DC. Western Lettuce.

L. pulchella (Pursh) Riddell (*L. tatarica* (L.) C. A. Mey. ssp. *pulchella (Pursh) Steb.*). Large Blue Lettuce.

L. saligna L. Willow-leaved Lettuce (34).

L. serriola L. Prickly Lettuce, Wild Lettuce.

L. serriola L.
    forma integrifolia Bogenhard.

LAGASCEA.

Lagascea decipiens Hemsl.

LAPSANA.

Lapsana communis L. Nipplewort (82).

LASTHENIA.

Lasthenia chrysostoma (F. & M.) Greene (*Baeria chrysostoma* F. & M.). Goldfields (212).

LAYIA.

Layia glandulosa (Hook.) H. & A. Tidy-tips.

LEONTODON. Hawkbit.

Leontodon rudicaulis (L.) Banks.

LESSINGIA.

Lessingia lemmoni Gray.

LEUCELENE.

Leucelene ericoides (Torr.) Greene (*Aster hirtifolius* Blake,
    A. arenosus (Heller) Blake). White Aster, Rose Heath (213).

LYGODESMIA. Skeleton Plant.

Lygodesmia arizonica Tomb (214).
L. grandiflora (Nutt.) T. & G.
L. spinosa Nutt. Thorny Skeleton Plant.

MACHAERANTHERA.

Machaeranthera aquifolia Woot. & Standl. (*Aster aquifolia* (Woot. &
    Standl.) Blake).
M. arida Turner & Horne (*Psilactis coulteri* Gray sensu K. & P.) (216).
M. arizonica Jackson & Johnston (included as a synonym within
    M. arida Turner & Horne). Hartman: Unpublished Ph.D.
    dissertation, Univ. of Texas, Austin, Texas (215).
M. bigelovii (Gray) Greene (*Aster bigelovii* Gray).
M. boltoniae (Greene) Turner & Horne (*Psilactis asteroides*
    Gray) (216).
M. canescens (Pursh) Gray (*Aster canescens* Pursh).
M. gracilis (Nutt.) Shinners (*Haplopappus gracilis* (Nutt.) Gray
    including *H. ravenii* Jackson). Hartman: Unpublished Ph.D.
    dissertation, Univ. of Texas, Austin, Texas.
M. grindelioides (Nutt.) Shinners
    var. *grindelioides (Haplopappus nuttallii* T. & G.).
M. juncea (Greene) Shinners (*Haplopappus junceus* Greene) (?).
M. linearis Greene (*Aster cichoriaceus* (Greene) Blake).
M. mexicana Turner & Horne (216).
M. mucronata Greene (*Aster adenolepis* Blake).
M. parviflora Gray (*Aster parvulus* Blake).
M. pinnatifida (Hook.) Shinners (217).
    ssp. *pinnatifida* var. *pinnatifida (Haplopappus spinulosus*
    (Pursh) DC. ssp. *spinulosis* sensu H. M. Hall; ssp. *typica*
    H. M. Hall (K. & P.); *H. spinulosus* (Pursh) DC. var. *turbinellus*
    (Rydb.) Blake).
    ssp. *pinnatifida* var. *chihuahuana* Turner & Hartman.
    ssp. *gooddingii* (A. Nels.) Turner & Hartman var. *gooddingii*
    (*H. spinulosus (*Pursh) DC. ssp. *gooddingii* (A. Nels.) Blake).
    ssp. *gooddingii* var. *paradoxa* Turner & Hartman.
M. tagetina Greene (*Aster tagetinus* (Greene) Blake).
M. tanacetifolia (H.B.K.) Nees (*Aster tanacetifolius* H.B.K.)
M. tephrodes (Gray) Greene (*Aster tephrodes* (Gray) Blake).

MADIA. Tarweed.

Madia glomerata Hook.

MALACOTHRIX.

Malacothrix californica DC.
    var. *glabrata* Eaton. Desert Dandelion.
M. clevelandi Gray. Cleveland Yellow Saucers.
M. coulteri Gray. Snakes Head.
M. fendleri Gray.
M. sonchoides (Nutt.) T. &. G. Yellow Saucers.
M. sonchoides (Nutt.) T. & G.
    var. *torreyi* (Gray) Williams.
M. sonorae Davis & Raven (50).
M. stebbinsii Davis & Raven (50).

MATRICARIA.

Matricaria chamomilla L. German Chamomille.
M. courrantiana DC.
M. matricarioides (Less.) Porter. Pineapple Weed.

MELAMPODIUM. Black Foot.

Melampodium leucanthum T. & G.
M. longicorne Gray.
M. strigosum Steussy (*M. hispidum* of auth. not H.B.K.) (218).

MICROSERIS.

Microseris heterocarpa (Nutt.) K. Chambers.
M. linearifolia (DC.) Schultz Bip. Silver Puffs.

MONOPTILON. Desert Star.

Monoptilon bellidiforme T. & G.
M. bellioides (Gray) H .M. Hall. Mohave Desert Star.

ODONTOTRICHUM. Indian Plantain.

Odontotrichum decompositum (A. Gray) Rydb. (*Cacalia decomposita* A. Gray) (219).

ONOPORDUM. Scotch Thistle.

Onopordum acanthium L. (28).

PALAFOXIA.

Palafoxia arida Turner & Morris (*P. linearis* (Cav.) Lag.). Spanish Needles (220).

PARTHENICE.

Parthenice mollis Gray.

PARTHENIUM.

Parthenium confertum Gray
    var. *lyratum* (Gray) Rollins.
P. incanum H.B.K.

PECTIS.

Pectis angustifolia Torr. Lemon-scented Pectis.
P. coulteri Harv. & Gray (?).
P. cylindrica (Fern.) Rydb.
P. filipes Harv. & Gray
    var. *subnuda* Fern. (221).
P. imberbis Gray.
P. linifolia L.
P. longipes Gray.
P. papposa Harv. & Gray. Chinchweed.
P. papposa Harv. & Gray
    var. *grandis* Keil (222).
P. prostrata Cav.
P. prostrata Cav.
    var. *urceolata* Fern.
P. rusbyi Gray (incl. *P. palmeri* Wats.) (221).

PENTZIA.

Pentzia incana (Thurb.) O. Kuntze (?).

PERICOME.

Pericome caudata Gray. Taperleaf Tailed Pericome.

PERITYLE (incl. LAPHAMIA). Rock Daisy (223,225,226).

**Perityle ajoensis* Todson (224).
P. ciliata (L. H. Dewey) Rydb.
P. cochisensis (Niles) Powell (Laphamia cochisensis Niles).
P. congesta (M. E. Jones) Shinners (*Laphamia congesta* Jones).
P. coronopifolia Gray.
P. emoryii Torr. (incl. var. *nuda* (Torr.) Gray). Emory Rock Daisy.
P. gilensis (Jones) Macbride (*Laphamia gilensis* Jones).
P. gilensis (Jones) Macbride
    var. *salensis* Powell.
P. gracilis (M. E. Jones) Rydb. (*Laphamia gracilis* M. E. Jones).
P. lemmoni (Gray) Macbride (*Laphamia lemmoni* Gray and *L.*
    dissecta Torr. sensu K. & P.

PERITYLE (incl. LAPHAMIA). Rocky Daisy (223,225,226). *(cont.)*

P. microglossa Benth. (Type of one synonomy i.e. var. *effusa*
A. Gray) (50).

P. saxicola (Eastw.) Shinners (*Laphamia saxicola* Eastw.).

P. tenella (Jones) Macbride (*Laphamia palmeri* A. Gray and
its var. *tenella* M. E. Jones).

PETRADORIA (227).

Petradoria pumila (Nutt.) Greene (*Solidago petradoria* Blake).
Rock Goldenrod.

P. pumila (Nutt.) Greene
ssp. *graminea* (Woot. & Standl.) L. C. Anderson (*Solidago
graminea* (Woot. & Standl.) Blake).

PEUCEPHYLLUM. Pigmy Cedar.

Peucephyllum schottii Gray. Desert Fir.

PINAROPAPPUS.

Pinaropappus roseus Less.
var. *foliosus* (Hellers) Shinners. Rock Lettuce.

PLATYSCHKUHRIA (228).

Platyschkuhria integrifolia (A. Gray) Rydb.
var. *oblongifolia* (A. Gray) Ellison (*Bahia oblongifolia* (A. Gray)
A. Gray).

PLEUROCORONIS.

Pleurocoronis pluriseta (Gray) King & Robinson (*Hofmeisteria
pluriseta* Gray). Arrow Leaf (229).

PLUCHEA. Marsh Fleabane, Stinkweed.

Pluchea purpurascens (Sw.) DC.
var *purpurascens* Godfrey. Canela.

PLUMMERA.

Plummera ambigens Blake.
P. floribunda Gray.

POROPHYLLUM.

Porophyllum gracile Benth. Odora.
P. ruderale (Jacq.) Cass.
ssp. *macrocephalum* (DC.) R. Roy Johns. (*P. macrocephalum*
DC.) (230).

170

PRENANTHELLA (231).

Prenanthella exigua (Gray) Rydb. (*Lygodesmia exigua* Gray).

PRIONOPSIS.

Prionopsis ciliata (Nutt.) Nutt. (*Haplopappus ciliatus* (Nutt.)
DC.) (56).

PSATHYROTES (232).

Psathyrotes annua (Nutt.) Gray. Fan Leaf.
P. pilifera Gray.
P. ramosissima (Torr.) Gray. Velvet Rosette.

PSILOSTROPHE. Paper Flower.

Psilostrophe cooperi (Gray) Greene.
P. sparsiflora (Gray) A. Nels.
P. tagetina (Nutt.) Greene.
P. tagetina (Nutt.) Greene
    var. *grandiflora* (Rydb.) Heiser.

PYRRHOPAPPUS. False Dandelion.

Pyrrhopappus multicaulis DC.

PYRROCOMA.

Pyrrocoma crocea (Gray) Greene (*Haplopappus croceus* Gray).
    var. *crocea*.

RAFINESQUIA. Plume Seed, Goatsbeard.

Rafinesquia californica Nutt. California Chicory.
R. neomexicana Gray. Desert Chicory, Desert Dandelion.

RATIBIDA. Mexican Hat.

Ratibida columnaris (Sims) D. Don.
R. columnaris (Sims) D. Don
    var. *pulcherrima* (DC.) D. Don.
R. tagetes (James) Barnhart. Prairie Coneflower.

RUDBECKIA. Brown-Eyed Susan.

Rudbeckia laciniata L.

SANVITALIA.

Sanvitalia abertii Gray.

SCHKUHRIA. Thread Leaf.

Schkuhria wislizenii Gray.
S. *wislizenii* Gray
 var. *frustrata* Blake.
 var. *wrightii* (Gray) Blake.

SENECIO. Groundsel, Ragwort.

Senecio actinella Greene.
S. *actinella* Greene
 var. *mogollonicus* (Greene) Greenm.
S. *arizonicus* Greene.
S. *bigelovii* Gray.
S. *cynthioides* Greene.
S. *douglasii* DC.
S. *douglasii* DC.
 var. *douglasii* (incl. S. *monoensis* Greene). Sand Wash
 Groundsel (233).
 var. *longilobus* (Benth.) L. Benson (S. *longilobus* Benth.).
 Thread Leaf Groundsel (233).
S. *eremophilus* Rich.
 var. *macdougalii* (Heller) Cronq. (S. *macdougalii* Heller) (234).
S. *franciscanus* Greene.
S. *hartianus* Heller.
S. *huachucanus* Gray.
S. *lemmoni* Gray.
S. *mohavensis* Gray. Mohave Groundsel.
S. *multicapitatus* Greenm.
S. *multilobatus* T. &. G. ex Gray (S. *lynceus* Greene, S. *millelobatus*
 Rydb. sensu K. & P., *stygius* Greene, S. *uintahensis* (A. Nels.)
 Greenm.) (235).
S. *mutabilis* Greenm. (236).
S. *mutabilis* Greenm. x S. *multilobatus* T. & G. ex Gray (236).
S. *mutabilis* Greenm. x S. *neomexicanus* Gray (236).
S. *neomexicanus* Gray.
S. *neomexicanus* Gray
 var. *griffithsii* Greenm.
S. *parryi* Gray. Mountain Groundsel.
S. *quercetorum* Greene.
S. *riddellii* T. & G. (233).
S. *salignus* DC.
S. *seemannii* Schultz Bip.
S. *spartioides* T. & G. Broom Groundsel.

S. vulgaris L. Common Groundsel.
S. werneriaefolius Gray.
S. werneriaefolius Gray
    var. *incertus* Greenm.
S. wootonii Greene.

SILYBUM. Milk Thistle.

Silybum marianum (L.) Gaertn.

SIMSIA. Bush Sunflower.

Simsia exaristata Gray.

SOLIDAGO. Goldenrod.

Solidago altissima L. Tall Goldenrod.
S. canadensis L.
    var. *gilvocanescens* Rydb. Canada Goldenrod.
S. decumbens Greene.
S. missouriensis Nutt.
S. missouriensis Nutt.
    var. *fasciculata* Holz.
S. multiradiata Ait.
S. nana Nutt.
S. occidentalis (Nutt.) T. & G. Western Goldenrod.
S. sparsiflora Gray.
S. wrightii Gray.
S. wrightii Gray
    var. *adenophora* Blake.
S. wrightii x *S. sparsiflora*.

SONCHUS. Sow Thistle.

Sonchus asper (L.) Hill. Spiny Sow Thistle, Achicoria Dulce.
S. oleraceus L. Annual Sow Thistle.

STENOTOPSIS.

Stenotopsis linearifolius (DC.) Rydb.
    var. *interior* (Coville) Macbr. (*Haplopappus linearifolius*
    DC. var. *interior* (Coville) M. E. Jones.).

STENOTUS.

Stenotus acaulis (Nutt.) Nutt. (*Haplopappus acaulis* (Nutt.)
    Gray var. *glabratus* DC.) (?).
S. armerioides Nutt. (*Haplopappus armerioides* (Nutt.) Gray).

STEPHANOMERIA. Wire Lettuce.

Stephanomeria exigua Nutt. Annual Mitra.
S. exigua Nutt.
    var. *pentachaeta* (D. C. Eaton) H. M. Hall.
S. parryi Gray. Parry Rock Pink.
S. pauciflora (Torr.) A. Nels. Desert Straw.
S. schottii Gray.
S. spinosa (Nutt.) Tomb (*Lygodesmia spinosa* Nutt.) (214).
S. tenuifolia (Torr.) H. M. Hall.
S. thurberi Gray.

STEVIA.

Stevia lemmoni Gray.
S. micrantha Lag.
S. plummerae Gray.
S. plummerae Gray
    var. *alba* Gray.
S. serrata Cav.
S. serrata Cav.
    var. *haplopappa* Robins.
    var. *ivaefolia* (Willd.) Robins.
S. viscida H.B.K.

STYLOCLINE.

Stylocline gnaphalioides Nutt. Everlasting Nest Straw.
S. micropoides Gray. Desert Nest Straw.

SYNTRICHOPAPPUS.

Syntrichopappus fremontii Gray. Fremont Xerasid.

TAGETES. Marigold.

Tagetes lemmoni Cav.
T. micrantha Cav. Licorice Marigold.

TANECETUM. Tansy.

*Tanacetum vulgare L. (34).

TARAXACUM. Dandelion.

Taraxacum laevigatum (Willd.) DC. Red-seeded Dandelion.
T. lyratum (Ledeb.) DC. Black-seeded Dandelion.
T. officinale Weber. Common Dandelion.

TESSARIA (237).

Tessaria sericea (Nutt.) Shinners (*Pluchea sericea* (Nutt.)
    Coville).

TETRADYMIA. Horse Brush.

Tetradymia axillaris A. Nels. Cotton Thorn.
T. canescens DC.
    var. *inermis* (Nutt.) Gray. Gray Felt Thorn, Black Sage.

THELESPERMA. Green Thread.

Thelesperma longipes Gray. Cota.
T. megapotamicum (Spreng.) Kuntze.
T. subnudum Gray. Navajo Tea.

TITHONIA.

Tithonia thurberi Gray.

TOWNSENDIA.

Townsendia annua Beaman.
T. exscapa (Richards) Porter.
T. formosa Greene.
T. incana Nutt. (*T. arizonica* Gray).
T. strigosa Nutt.

TRAGOPOGON. Goats Beard.

Tragopogon dubius Scop.
**T. minus* Ownbey (238).
**T. miscellus* Ownbey (238).
T. porrifolius L. Salsify, Oyster Plant.
T. pratense L.

TRICHOPTILIUM.

Trichoptilium incisum Gray. Yellow Head.

TRIXIS.

Trixis californica Kellogg.

VANCLEVEA.

Vanclevea stylosa (Eastw.) Greene.

VERBESINA. Crown Beard.

Verbesina encelioides (Cav.) Benth. & Hook. Yellowtop, Cowpen
    Daisy.
V. encelioides (Cav.) Benth. & Hook.
    var. *exauriculata* Robins. & Greenm. Crownbeard.
V. longifolia Gray.
V. rothrockii Robins. & Greenm.

VIGUIERA. Golden Eye.

Viguiera annua (Jones) Blake. Annual Goldeneye.
V. ciliata (Robins. & Greenm.) Blake.
V. cordifolia Gray.
V. deltoidea Gray.
    var. *parishii* (Greene) Vasey & Rose. Parish Viguiera.
V. dentata (Cav.) Spreng.
V. dentata (Cav.) Spreng.
    var. *lancifolia* Blake.
V. longifolia (Robins. & Greenm.) Blake.
V. multiflora (Nutt.) Blake.
V. multiflora (Nutt.) Blake
    var. *nevadensis* (A. Nels.) Blake. Nevada Viguiera.
V. ovalis Blake.

WYETHIA. Mules Ears.

Wyethia arizonica Gray.
W. scabra Hook.
W. scabra Hook.
    var. *attenuata* W. Weber.
    var. *canescens* W. Weber.

XANTHIUM. Cocklebur.

Xanthium spinosum L. Spiny Cocklebur.
X. strumarium L. (*X saccharatum* Wallr.) Common Cocklebur, Abrojo.

XANTHOCEPHALUM. Broomweed, Snakeweed.

Xanthocephalum wrightii Gray.

XYLORHIZA (239).

Xylorhiza tortifolia (T. & G.) Greene (*Machaeranthera tortifolia* (Gray) Cronq. & Keck). Mohave Aster, Desert Aster.
X. tortifolia (T. & G.) Greene
    var. *imberbis* (Cronq.) T. J. Watson.

ZALUZANIA.

Zaluzania grayana Robins. & Greenm.

ZEXMENIA.

Zexmenia podocephala Gray.

ZINNIA (240).

Zinnia acerosa (DC.) Gray (*Z. pumila* Gray).
Z. grandiflora Nutt. Prairie Zinnia.
Z. peruviana (L.) L. (*Z. multiflora* L.)

LITERATURE CITED

1. Kearney, Thomas H.; Peebles, Robert H. and Collaborators. Arizona Flora with Supplement by Howell, John Thomas; McClintock, Elizabeth; and Collaborators. Univ. Calif. Press, Berkeley and Los Angeles, Calif., 1960.

2. Willis, J. C. A Dictionary of the Flowering Plants and Ferns. Eighth Edition rev. by H. K. Airy Shaw, Cambridge, Great Britain, 1973.

3. Benson, Lyman. The Cacti of Arizona. Third Edition. Univ. Ariz. Press, Tucson, Ariz., 1969.

4. Weatherby, Una F. The English Names of North American Ferns. Amer. Fern Journ. 42: 134–151, 1952.

5. Hitchcock, A. S. Manual of Grasses of the United States. Rev. by Chase, Agnes. U.S. Dept. of Agr. Misc. Pub. 200, Wash., D.C., 1951.

6. Little, Elbert L. Jr. Check List of Native and Naturalized Trees of the United States, (incl. Alaska). Forest Svc. Agr. Hdbk. No. 41, Wash. D.C., 1953.

7. Parker, Kitty F. An Illustrated Guide to Arizona Weeds. Univ. Ariz. Press, Tucson, Ariz., 1972.

8. Jaeger, Edmund C. Desert Wild Flowers. Stanford Univ. Press, Stanford, Calif. Rev. Edit., 1941.

9. Cory, V. L. and Parks, H. B. Catalogue of the Flora of Texas. Agr. & Mech. College of Texas, College Station, Texas, 1937.

10. Reeves, Timothy. Range Extensions for Lower Vascular Plants in Arizona. Madrono 23: 454–455, 1976.

11. Mason, Charles T. Jr. A New Family of Vascular Plants (Psilotaceae) for Arizona. Madrono 19: 224, 1968.

12. Reeves, Timothy. The Genus Botrychium (Ophioglossaceae) in Arizona. Amer. Fern Journ. 67: 33–39, 1977.

13. Pinkava, D. J.; Lehto, Elinor; Reeves, Tim; and Sundell, Eric. Plants New to Arizona Flora. Journ. Ariz. Acad. 10: 146, 1975.

14. Cronquist, A.; Holmgren, A. H.; Holmgren, N. H.; and Reveal, J. L. Intermountain Flora, Vol. 1, 1972. N.Y. Botanical Garden, New York, N.Y.

15. Tryon, Rolla, A Revision of the American Species of Notholaena. Contr. Gray Herb. 179: 1–106, 1956.

16. Hevly, Richard H. Studies of the Sinuous Cloak-Fern (Notholaena sinuata) Complex. Journ. Ariz. Acad. 3: 205–208, 1965.

17. Tryon, Alice F. A. Revision of the Fern Genus Pellaea Section Pellaea. Ann. Mo. Bot. Gard. 44: 125–193, 1957.

18. Smith, Alan R. Systematics of the Neotropical Species of Thelypteris Section Cyclosorus. Univ. Cal. Pub. Bot. 59: 1–143, 1971.

19. Vasek, Frank C. The Distribution and Taxonomy of Three Western Junipers. Brittonia 18: 350–372, 1966.

20. Franco, J. D. A Taxonomy of the Common Juniper. Bol. Soc. Brot. 2. 36: 101–120, 1962.

21. Cutler, Hugh Carson. Monograph of the North American Species of the Genus Ephedra. Ann. Mo. Bot. Gard. 26: 373–428, 1939.

22. Mason, Charles T. Jr., and Hevly, Richard H. Additions to the Aquatic Flora of Arizona. Madrono 16: 32, 1961.

23. Hevly, Richard H. Notes on Arizona Plants. Plateau 34: 135–136, 1962.

24. St. John, Harold. Monograph of the Genus Elodea, Summary. Rhod. 67: 155–180, 1965.

25. Correll, Donovan S., and Correll, Helen B. Aquatic and Wetland Plants of Southwestern United States. Environmental Protection Agency, Wash. D.C., 1972.

26. Cronquist, A.; Holmgren, A. H.; Holmgren, N. H.; Holmgren, P. K.; and Reveal, J. L. Intermountain Flora, Vol. 6, 1977. N.Y. Botanical Garden, New York, N.Y.

27. Mason, Herbert L. A. Flora of the Marshes of California. Univ. Calif. Press, Berkeley, Calif., 1957.

28. Pinkava, D. J. and Lehto, Elinor. Plants New to Arizona Flora. Journ. Ariz. Acad. 5: 27, 1968.

29. Howell, J. T. Remarks on Triglochin concinna. Leafl. West. Bot. 5: 13–19, 1947.

30. Correll, D. S., and Johnston, M. C. Manual of the Vascular Plants of Texas. Texas Research Foundation, Renner, Texas, 1970.

31. Gould, Frank W. The Grasses of Texas. Texas A. & M. Univ. Press, College Station, Texas, 1975.

32. Gould, F. W., and Kapadia, Z. J. Biosystematic Studies in the Bouteloua curtipendula Complex 2. Taxonomy. Brittonia 16: 182–207, 1964.

33. Gould, Frank W. Taxonomy of the Bouteloua repens Complex. Brittonia 21: 261–274, 1969.

34. McDougall, W. B. Seed Plants of Northern Arizona. Museum of Northern Arizona, Flagstaff, Ariz., 1973.

35. DeLisle, D. G. Taxonomy and Distribution of the Genus Cenchrus. Iowa State Coll. Journ. Sci. 37: 259–351, 1963.

36. Yates, H. O. Revision of Grasses Traditionally Referred to Uniola. Chasmanthium. Southw. Nat. 11: 415–455, 1966.

37. Lehr, J. Harry. Chloris cucullata in Arizona. Journ. Ariz. Acad. 7: 51, 1972.

38. Gould, Frank W. Nomenclatural Changes in the Poaceae. Brittonia 26: 59–60, 1974.

39. Ebinger, John E. Validity of the Grass Genus Digitaria adscendens. Brittonia 14: 248–253, 1962.

40. Beetle, Alan A. The North American Variations of Distichlis spicata. Bull. Torr. Club 70: 638–650, 1943.

41. Gould, Frank W. Grasses of Southwestern United States. Univ. Ariz. Press, Tucson, Ariz., 1951.

42. Pinkava, Donald J.; Lehto, Elinor; and Keil, David. Plants New to Arizona Flora. Journ. Ariz. Acad. 6: 134, 1970.

43. Tateoka, Tuguo. A Biosystematic Study of Tridens (Gramineae). Amer. Journ. Bot. 48: 565–573, 1961.

44. Munz, Philip A. A Flora of Southern California. Univ. Calif. Press, Berkeley, Calif., 1974.

45. Banks, Donald J. Taxonomy of Paspalum setaceum (Gramineae). Sida 2: 269–284, 1966.

46. Howell, J. T. Harding Grass in Arizona. Leafl. West. Bot. 10: 41, 1963.

47. Clayton, W. D. The Correct Name of the Common Reed. Taxon 17: 157–158, 1968.

48. Rubtzoff, Peter. Interesting Records in Polypogon from California and Arizona. Leafl. West. Bot. 10: 119–120, 1964.

49. Gould, Frank W. The Grass Genus Andropogon in the United States. Brittonia 19: 70–76, 1967.

50. Shreve, Forrest, and Wiggins, Ira. L. Vegetation and Flora of the Sonoran Desert. 2 vol. Stanford Univ. Press, Stanford, Calif., 1964.

51. Rominger, J. M. Taxonomy of Setaria (Gramineae) in North America. Univ. Ill. Biol. Mono. 29: 1–132, 1962.

52. Erdman, K. S. Taxonomy of the Genus Sphenopholis. Iowa State Coll. Journ. Sci. 30; 259–336, 1965.

53. Shinners, Lloyd H. Notes on North Texas Grasses. Rhod. 56: 28, 1954.

54. Lonard, R. I., and Gould, F. W. The North American Species of Vulpia (Gramineae). Madrono 22: 217–230, 1974.

55. Pinkava, Donald J.; Lehto, Elinor; and Higgins, Larry C. Plants New to Arizona Flora. 4. Journ. Ariz. Acad. 2: 51, 1972.

56. Mason, Charles T., Jr. Notes on the Flora of Arizona. 5. Journ. Ariz. Acad. 6: 189, 1971.

57. Hermann, Frederick J. Manual of the Carices of the Rocky Mountains and Colorado Basin. Agr. Hdbk. 374, U.S. Dept. of Agr., Wash., D.C., 1970.

58. Ibid. Notes on Rocky Mountain Carices. Rhod. 70: 419, 1968.

59. Johnston, Marshall C. Cyperus huarmensis (H.B.K.) M. C. Johnst. Comb. Nov. Southw. Nat. 11: 124–125, 1966.

60. Daubs, E. H. A Monograph of Lemnaceae. Univ. Ill. Biol. Mono. 34: 1–118, 1965.

61. Hermann, F. J. Manual of the Rushes (Juncus ssp.) of the Rocky Mountains and Colorado Basin. U.S.D.A. Forest Svc. Tech. Rpt. RM 18, Rocky Mtn. Forest and Range Experi. Sta., Fort Collins, Colo., 1975.

62. Gentry, Howard Scott. Two New Agaves in Arizona. Cac. & Succ. Journ. 42: 222–228, 1970.

63. Crawford, Daniel J. A. Morphological and Chemical Study of Populus acuminata Rybd. Brittonia 26: 74–89, 1974.

64. Dorn, Robert D. Willows of the Rocky Mountain States. Rhod. 79: 390–429, 1977.

65. Tucker, John M. Studies in the Quercus undulata Complex. 1. A Preliminary Statement. Amer. Journ. Bot. 48: 202–208, 1961.

66. Muller, Cornelius H. The Oaks of Texas. Contr. Texas Res. Found. 1: 21–323, 1951. Renner, Texas.

67. Tucker, John M.; Cottam, Walter P.; and Drobnick, Rudy. Studies in the Quercus undulata Complex. 2. The Contribution of Quercus turbinella. Amer. Journ. Bot. 48, 329–339, 1961.

68. Howell, J. T. Concerning Two Asiatic Elms. Leafl. West Bot. 10: 328–329, 1966.

69. Hinton, B. D. Parietaria hespera (Urticaceae), a New Species of the Southwestern United States. Sida 3: 293–297, 1969.

70. Hawksworth, Frank G., and Wiens, Delbert. Biology and Classification of Dwarf Mistletoes (Arceuthobium). Agr. Hdbk. 401, U.S. Dept. of Agr., Wash., D.C., 1972.

71. Ibid. New Taxa and Nomenclatural Changes in Arceuthobium (Viscaceae). Brittonia 22: 265–268, 1970.

72. Ibid. A New Species of Arceuthobium from Arizona. Ibid. 16: 54–57, 1964.

73. Wiens, Delbert. Revision of Acatophyllus Species of Phoradendron. Ibid. 16: 11–54, 1964.

74. Reveal, James L. Eriogonum (Polygonaceae) of Arizona and New Mexico. Phytologia 34:409–484,1976.

75. Ibid. A New Perennial Buckwheat (Eriogonum, Polygonaceae) from Southwestern Arizona. Journ. Ariz. Acad. 5: 222–225, 1969.

76. Ibid. Two Shrubby Novelties in Eriogonum (Polygonaceae) from the Deserts of Utah and Arizona. Brittonia 26: 90–94, 1974.

77. Coolidge, Jerrold. A New Species of Polygonum (P. triandrous (Polygonaceae). Madrono 20: 226–269, 1970.

78. Rechinger, K. H., Jr. The North American Species of Rumex. Field Museum of Nat. Hist. Bot. Series 17: No. 1, 1937.

79. Reveal, J. L. and Ertler, B. J. Re-establishment of the Genus Stenogonum (Polygonaceae). Great Basin Nat. 36: 272–280, 1976.

80. Hall, H. M., and Clements, F. E. The Phylogenetic Method in Taxonomy. Genus Atriplex. Carn. Inst. Wash. Pub. 326: 235–346, 1923, Wash., D.C.

81. Crawford, Daniel J. On the Relationship of Chenopodium flabellifolium and C. inamoenum. Madrono 24: 63–64, 1977.

82. Pinkava, D. J.; Lehto, Elinor; and Keil, David. Plants New to Arizona Flora. 2. Journ. Ariz. Acad. 5: 226, 1969.

83. Beatley, Janice C. Russian-Thistle (Salsola) Species in Western United States. Journ. Range Mgmt. 26: 225–226, 1973.

84. McNeill, J.; Bassett, I. J.; and Crompton, C. W. Suaeda calceoliformis, the Correct Name for Suaeda depressa Auct. Rhod. 79: 133–137, 1977.

85. Galloway, Leo A. Systematics of the North American Desert Species of Abronia and Tripterocalyx (Nyctaginaceae). Brittonia 27: 328–345, 1975.

86. Fowler, B. A., and Turner, B. L. Taxonomy of Selinocarpus and Ammocodon (Nyctaginaceae). Phytologia 37: 177–208, 1977.

87. Lane, Meredith and Keil, David J. Glinus radiatus (Aizoaceae) Chromosome Count and Range Extension to Arizona. Madrono 23: 457, 1976.

88. Miller, John M. Variation in Populations of Claytonia perfoliata (Portulacaceae). Sys. Bot. 1: 20–34, 1976.

89. Duke, James A. Preliminary Revision of the Genus Drymaria. Ann. Mo. Bot. Gard. 48: 173–268, 1961.

90. McDougall, W. B. Notes on Northern Arizona Plants. Plateau 37: 107, 1965.

91. Beal, Ernest O. Taxonomic Revision of Genus Nuphar Sm. of North America and Europe. Journ. Elisha Mitchell Sci. Soc. 72: 317–346, 1956.

92. Mason, C. T., Jr. Notes on the Flora of Arizona. 3. Madrono 17: 236, 1964.

93. Lewis, H. and Epling, C. A Taxonomic Study of California Delphiniums. Brittonia 8: 1–22, 1954.

94. Pitman, Marsh. Coronopus and Sherardia in Arizona. Leafl. West. Bot. 9: 256, 1962.

95. Rollins, Reed C., and Shaw, Elizabeth A. The Genus Lesquerella (Cruciferae) in North America. Harvard Univ. Press, Cambridge, Mass., 1973.

96. Lehr, J. Harry. Some Adventives New to the Flora of Arizona. Journ. Ariz. Acad. 9: 109, 1974.

97. Al-Shehbaz, Ihsan A. The Biosystematics of the Genus Thelypodium (Crucifera). Contr. Gray Herb. 204: 3–148, 1973.

98. Holmgren, Patricia K. A Biosystematic Study of North American Thlaspi montanum and its Allies. Mem. N.Y. Bot. Gard. 21: 1–106, 1971.

99. Rollins, Reed C. Chromosome Numbers of Cruciferae. Contr. Gray Herb. 197: 43–65, 1966.

100. Iltis, Hugh H. Studies in the Capparidaceae 5. Capparidaceae of New Mexico. Southw. Nat. 3: 133–144, 1958.

101. Ibid. Studies in the Capparidaceae 8. Polanisia dodecandra (L.) DC. Rhod. 68: 41–47, 1966.

102. Clausen, R. T. A Reinterpretation of Sedum stenopetalum and Sedum lanceolatum. Journ. Cac. & Succ. Soc. Amer. 20: 143–146, 1948.

103. Jones, George Neville. American Species of Amelanchier. Univ. Ill. Biol. Mono. 20: 1–100, 1946.

104. Mason, C. T. Jr. Apacheria chiricahuensis: A New Genus and Species from Arizona. Madrono 23: 105–108, 1975.

105. Martin, Floyd L. A Revision of Cercocarpus. Brittonia 7: 91–111, 1950.

106. Wells, Phillip V., and Johnson R. Roy. Vauquelinia pauciflora (Rosaceae) from Guadalupe Canyon, Arizona: a Species of
Trees Newly Reported for the United States. Southw. Nat. 9:
151–154, 1964.

107. Isely, Duane. Leguminosae of the U.S. 1. Mimosoideae.
Mem. N.Y. Bot. Gard. 25: 1973.

108. Burkart, Arturo. A Monograph of the Genus Prosopis.
Journ. Arn. Arb. 57: 219–249, 1976.

109. Eifert, Imre J. New Combinations in Hoffmanseggia Cav.
and Caesalpinia L. Sida 5: 43–44, 1972.

110. Barneby, Rupert C. Atlas of North American Astragalus.
Mem. N.Y. Bot. Gard. 13: 2 Parts, 1964.

111. Welsh, Stanley L. and Thorne, Kaye H. Plants of Arizona:
A New Species of Astragalus from the Kaibab Plateau. Great
Basin Nat. 37: No. 1, 1977.

112. Fearing, Olin S. A Cytotaxonomic Study of the Genus
Cologania and its Relationship to Amphicarpa (Leguminosae-
Papilionoideae), a Doctoral Dissertation Done and Filed at the
University of Texas, 1959.

113. Barneby, R. C. A Synopsis of Errazurizia. Leafl. West.
Bot. 9: 209–214, 1962.

114. Hess, Lloyd W. and Dunn, David B. Nomenclature of the
Lupinus argenteus and L. caudatus Complexes. Rhod. 72: 110–
114, 1970.

115. Dunn, David B.; Christian, James A. and Dziekanowski,
Chester T. Nomenclature of the California Lupinus concinnus –
L. sparsiflorus Complex. Aliso 6: 45–50, 1966.

116. Wemple, Don K. Revision of the Genus Petalostemon
(Leguminosae). Iowa State Journ. Sci. 45: 1–102, 1970.

117. Hermann, F. J. Vetches of the United States – Native,
Naturalized and Cultivated. U.S. Dept. Agr., Agr. Hdbk. No. 168,
Wash., D.C., 1960.

118. Eiten, George. Taxonomic and Regional Variations of
Oxalis Section Corniculatae. Amer. Mid. Nat. 69: No. 2, 1963.

119. Denton, Melinda F. A. Monograph of Oxalis. Mich. State
Univ. Biol. Series 4: No. 10, 1973.

120. Rogers, C. M. Yellow-flowered Species of Linum in Central America and Western North America. Brittonia 20: 107–135,
1968.

121. Bailey, Virginia Long. Revision of the Genus Ptelea (Ruta-
ceae). Ibid. 14: 1–45, 1962.

122. Moran, Reid and Felger, Richard. Castela polyandra,
a New Species in a New Section; Union of Holacantha with Cas-

183

tela (Simaroubaceae). Trans. San Diego Museum Nat. Hist. 15: 3–40, 1968.

123. Miller, Kim I., and Webster, Grady L. A Preliminary Revision of Tragia (Euphorbiaceae) in the United States. Rhod. 69: 241–305, 1967.

124. Johnston, Marshall C. Revision of Condalia including Microrhamnus (Rhamnaceae). Brittonia 14: 332–368, 1962.

125. Johnston, Laverne A. Revision of the Rhamnus serrata Complex. Sida 6: 67–80, 1975.

126. Brizicky, G. K. Herisantia, Bogenhardia, and Gayoides (Malvaceae). Journ. Arn. Arb. 49: 278–279, 1968.

127. Fryxell, Paul A. The North American Malvellas (Malvaceae). Southw. Nat. 19: 97–103, 1974.

128. Clement, I. D. Studies in Sida (Malvaceae). 1. A Review of the Genus and Monograph of the Sections Malacroideae, Physalodes, Pseudo-malvastrum, Incanifolia, Oligandrae, Pseudonapaea, Hookeria and Steninda. Contr. Gray Herb. 180: 1–91, 1957.

129. Shinners, Lloyd H. Three New Varietal Names in Sphaeralcea (Malvaceae). Sida 1: 384–385, 1964.

130. Cristobal, Carmen L. Revision del Genero Ayenia (Sterculiaceae). Opera Lilloana 4: 230p, 1960. Tucuman, Republica Argentina.

131. Hevly, Richard H. New Species for Arizona. Plateau 33: 116–117, 1961.

132. Clausen, J. New Combinations in Western North American Violets. Madrono 17: 295, 1964.

133. Russell, Norman H. Viola palustris L. in Arizona. Rhod. 65: 49, 1963.

134. Earle, Hubert J. Cochiseia Earle, genus novum. Cochiseia robbinsorum Earle, species nova. Saguaroland Bull. 30: No. 6, 1976.

135. Zimmerman, Allen D., and Dale, A. A Revision of the United States Taxa of the Mammillaria wrightii Complex with Remarks upon the Northern Mexican Populations. Part 2. Cac. & Succ. Journ. 49: 51–62, 1977.

136. Pinkava, D. J.; McLeod, M. G.; McGill, L. A.; and Brown, R. C. Chromosome Numbers in some Cacti of Western North America — 2. Brittonia 25: 2–9, 1973.

137. Benson, Lyman. New Taxa and Nomenclatural Changes in the Cactaceae. Cac. & Succ. Journ. 46: 79–81, 1974.

138. Grant, Verne and Grant, Karen A. Natural Hybridization Between the Cholla Cactus Species Opuntia spinosior and Opuntia versicolor. Proc. Nat. Acad. U.S. 68: 1993–1995, 1971.

139. Woodruff, D. and Benson, L. Changes of Status in Sclero-cactus. Cac. & Succ. Journ. 48: 131–134, 1976.

140. Munz, Phillip A. Onagraceae. N. A. Flora, Series 2, Part 5. N.Y. Bot. Gard., 1965.

141. Towner, Howard F. The Biosystematics of Calylophus (Onagraceae). Ann. Mo. Bot. Gard. 64: 48–120, 1977.

142. Raven, Peter H. A Revision of the Genus Camissonia (Onagraceae). Contr. U.S. Nat. Herb. 37: Parts, 1969.

143. Mosquin, Theodore. A New Taxonomy for Epilobium angustifolium (Onagraceae). Brittonia 18: 167–191, 1966.

144. Seavey, Steven R.; Wright, Phillip; Raven, Peter H. A Comparison of Epilobium minutum and E. foliosum (Onagraceae). Madrono 24: 6–12, 1977.

145. Raven, Peter H. and Gregory, David P. A Revision of the Genus Gaura (Onagraceae). Mem. Torr. Bot. Club 23: No. 1, 1972.

146. Lewis, Harlan D. and Szweykowski, Jerzy. The Genus Gayophytum (Onagraceae). Brittonia 16: 343–391, 1964.

147. Raven, Peter H. and Parnell, Dennis R. Two New Species and Some Nomenclatural Changes in Oenothera subg. Hartman-nia (Onagraceae). Madrono 20: 146–149, 1970.

148. Lehr, J. Harry. Some Additions to the Flora of Arizona. Journ. Ariz. Acad. 11: No. 1, 1976.

149. Mason, Charles J., Jr., and Niles, Wesley E. Notes of the Flora of Arizona. 4. Madrono 19: No. 5, 1968.

150. Theobald, William L. The Lomatium dasycarpum-mohav-ense-foeniculaceum Complex (Umbelliferae). Brittonia 18: 1–18, 1966.

151. Chuang, T. and Constance, L. A Systematic Study of Perideridia. Univ. Cal. Pub. Bot. 55: 1–74, 1969.

152. McDougal, W. B. and Stockert, John. New Plants for Arizona from Grand Canyon National Park and Marble Canyon. Plateau 39: 102–103, 1966.

153. Shinners, Lloyd H. Texas Asclepiadaceae other than Asclepias. Sida 1: 358–367, 1964.

154. Brummitt, R. K. New Combinations in North American Calystegia. Ann. Mo. Bot. Gard. 52: 214–216, 1965.

155. Grant, Alva and Grant, V. The Genus Allophyllum. Aliso 3: 93–110, 1955.

156. Grant, V. A Synopsis of Ipomopsis. Ibid. 3: 351–362, 1956.

157. Patterson, Robert. A Revision of Linanthus Sect. Sipho-nella (Polemoniaceae). Madrono 24: 36–48, 1977.

185

158. Atwood, Duane N. A Revision of the Phacelia crenulata Group (Hydrophyllaceae) for North America. Great Basin Nat. 35: 127–190, 1975.

159. Gillett, George W. A Systematic Treatment of the Phacelia franklinii Group. Rhod. 62: 205–222, 1960.

160. Higgins, Larry C. A Revision of Cryptantha Subgenus Oreocarpa. Brigham Young Univ. Sci. Bull. Bio. Series 13: No. 4, 1971.

161. Higgins, Larry C. Cryptantha atwoodii (Boraginaceae) A New Species from Arizona. Southw. Nat. 19: 127–130, 1974.

162. Richardson, A. Reinstatement of the Genus Tiquilia (Boraginaceae, Ehretioideae) and Description of Four New Species. Sida 6: 235–240, 1976.

163. Ibid. Monograph of the Genus Tiquilia (Coldenia, sensu lato), Boraginaceae: Ehretioideae. Rhod. 79: 467–572, 1977.

164. Shinners, Lloyd H. Verbena pulchella Sweet var. gracilior (Troncoso) Shinners, comb. nov. (Verbenaceae). Sida 2: 266, 1966.

165. Averett, John E. Biosystematic Study of Chamaesaracha (Solanaceae). Rhod. 75: 325–365, 1973.

166. Wells, Philip V. Variation in Section Trigonophylla of Nicotiana. Madrono 15: 148–151, 1960.

167. Sandwith, N. Y. The Identity of Saracha acutifolia Miers. Kew Bull. 14: 232, 1960.

168. Waterfall, U. T. A Taxonomic Study of the Genus Physalis in North America North of Mexico. Rhod. 60: 106–114, 152–173, 1958.

169. Shinners, Lloyd H. Salpichroa origanifolia instead of S. rhomboidea (Solanaceae). Leafl. West Bot. 9: 257–259, 1962.

170. Holmgren, Noel H. Five New Species of Castilleja (Scrophulariaceae) from the Intermountain Region. Bull Torr. Club 100: 83–93, 1973.

171. Ibid. Four New Species of Mexican Castilleja, (Subgenus Castilleja, Scrophulariaceae) and Their Relatives. Brittonia 28: 195–208, 1976.

172. Straw, Richard M. Keckiella: New Name for Keckia Straw (Scrophulariaceae). Ibid. 19: 203–204, 1967.

173. Carr, Gerald D. Taxonomy of Pedicularis parryi (Scrophulariaceae). Ibid. 23: 280–291, 1971.

174. Crosswhite, Frank S. Hybridization of Penstemon barbatus (Scrophulariaceae) of Section Elmigera with Species of a Section Habroanthus, Southw. Nat. 10: 234–237, 1965.

175. Ibid. Revision of Penstemon Section Chamaeleon (Scrophulariaceae). Sida 2: 339–346, 1966.

176. Haynes, Robert R. Notes: Ibid. 3: 347, 1969.

177. Dempster, Lauramay T. Galium mexicanum (Rubiaceae) of Central America and Western North America. Madrono 23: 378–386, 1976.

178. Dempster, Lauramay T. and Ehrendorfer, Friedrich. Evolution of the Galium multiflorum Complex in Western North America. 2. Critical Taxonomic Revision. Brittonia 17: 289–334, 1965.

179. Bemis, W. P. & Whitaker, Thomas W. Natural Hybridization Between Cucurbita digitata and C. palmata. Madrono 18: 39–46, 1965.

180. Abrams, Leroy and Ferris, Roxana Stinchfield. Illustrated Flora of the Pacific States 4: 390–391, 1960. Stanford Univ. Press, Stanford, Calif.

181. Reveal, James L., and King, Robert M. Re-establishment of Acourtia D. Don (Asteraceae). Phytologia 27: 228–232, 1973.

182. King, R. M. and Robinson, H. New Combinations in Ageratina. Ibid. 19: 208–229, 1970.

183. Payne, Willard W. A Re-evaluation of the Genus Ambrosia (Compositae). Journ. Arn. Arb. 45: 401–438, 1964.

184. Maguire, B. A. A Monograph of the Genus Arnica. Brittonia 4: 386–510, 1943.

185. Hall, H. M. and Clements, F. E. The Phylogenetic Method in Taxonomy. Genus Artemisia. Carn. Inst. Wash. Pub. 326: 31–156, 1923.

186. Cuatrecasas, Jose. Notas Adicionales Taxonomicas y Carologicas, Sobre Baccharis. Revista Acad. Colomb. 13: 210–226, 1968.

187. Ellison, William L. A Systematic Study of the Genus Bahia (Compositae). Rhod. 66: 66–86, 177–215, 281–311, 1964.

188. Shinners, Lloyd H. Kuhnia L. Transferred to Brickellia Ell. (Compositae). Sida 4: 274, 1971.

189. Anderson, Loran C. Taxonomic Notes on the Chrysothamnus viscidiflorus Complex (Astereae, Compositae). Madrono 17: 222–227, 1964.

190. Ibid. Cytotaxonomic Studies in Chrysothamnus (Astereae, Compositae). Amer. Journ. Bot. 53: 204–212, 1966.

191. Cronquist, A. The Separation of Erigeron from Conyza. Bull. Torr. Club 70: 629–632, 1943.

192. Sharsmith, Helen K. The Native California Species of the Genus Coreopsis. Madrono 4: 209–231, 1938.

193. Smith, Edwin B. and Parker, Hampton M. A Biosystematic Study of Coreopsis tinctoria and C. cardaminefolia (Compositae). Brittonia 23: 161–176, 1971.

187

194. Earle, W. Hubert. A New Plant for Arizona, Dimorpho-theca aurantiaca. Saguaroland Bull. 27: 56–57, 1973.

195. Bierner, Mark W. A Systematic Study of Dugaldia (Compositae). Brittonia 26: 385–392, 1974.

196. Strother, J. L. Systematics of Dyssodia Cav. Univ. Cal. Pub. Bot. 48: 1–88, 1969.

197. Kyhos, Donald W. Natural Hybridization Between Encelia and Geraea (Compositae) and Some Related Experimental Investigations. Madrono 19: 33–43, 1967.

198. Blake, S. F. A Revision of Encelia and Some Related Genera. Proc. Am. Acad. 49: 358–376, 1913.

199. Urbatsch, Lowell D. Systematics of the Ericameria cuneata Complex (Compositae: Asterae). Madrono 23: 338–345, 1976.

200. Theroux, Michael E.; Pinkava, Donald J. and Keil, David J. A New Species of Flaveria (Compositae: Flaverinae) from Grand Canyon, Arizona. Ibid. 24: 13–17, 1977.

201. Canne, Judith M. A Revision of the Genus Galinsoga (Compositae: Helenieae). Univ. Kan. Sci. Bull. 48: 225–267, 1969.

202. Harrington, H. D. Manual of the Plants of Colorado, 624, 1954. Denver, Colo.

203. Ruffin, John. A New Combination in Grindelia (Compositae-Astereae). Rhod. 79: 583–585, 1977.

204. Solbrig, O. T. Cytotaxonomic and Evolutionary Studies in· the North American Species of Gutierrezia (Compositae). Contr. Gray Herb. 188: 3–86, 1960.

205. Ibid. Note on Gymnosperma glutinosum (Compositae-Astereae). Leafl. West. Bot. 9: 147–150, 1961.

206. Heiser, Charles B. Jr. with Smith, Dale M.; Clevenger, Sarah B.; Martin, William C. Jr. The North American Sunflowers (Helianthus). Mem. Torr. Bot. Club 22: No. 3, 1969.

207. Harms, Vernon L. Cytogenetic Evidence Supporting the Merger of Heterotheca and Chrysopsis (Compositae). Brittonia 17: 11–16, 1965.

208. Ibid. Nomenclatural Change and Taxonomic Notes on Heterotheca including Chrysopsis in Texas and Adjacent States. Wrightia 4: 8–20, 1968.

209. Peterson, Kathleen M. and Payne, Willard W. On the Correct Name for the Appressed-winged Variety of Hymenoclea salsola (Compositae:Ambrosieae). Brittonia 26: 397, 1974.

210. Ibid. The Genus Hymenoclea (Compositae:Ambrosieae). Ibid. 25: 243–256, 1973.

211. Jackson, R. C. A Revision of the Genus Iva L. Univ. Kan. Sci. Bull. 41: 793–876, 1960.

212. Ornduff, R. A. A Biosystematic Survey of the Goldfield Genus Lasthenia. Univ. Cal. Pub. Bot. 40: 1–92, 1966.

213. Shinners, L. H. Revision of the Genus Leucelene. Wrightia 1: 82–89, 1946.

214. Tomb, Andrew Spencer. Novelties in Lygodesmia and Stephanomeria (Compositae-Cichorieae). Sida 3: 530–532, 1970.

215. Jackson, R. C. and Johnson, R. R. A New Species of Machaeranthera Section Psilactis (M. arizonica). Rhod. 69: 476–480, 1967.

216. Turner, B. L. and Horne, David. Taxonomy of Machaeranthera Sect. Psilactis (Compositae-Astereae). Brittonia 16: 316–331, 1964.

217. Turner, B. L. and Hartman, R. Infraspecific Categories of Machaeranthera pinnatifida (Compositae). Wrightia 5: 308–315, 1976.

218. Steussy, Ted F. Revision of the Genus Melampodium (Compositae-Heliantheae). Rhod. 74: 1–70, 1972.

219. Pippen, Richard W. Mexican "Cacalioid" Genera Allied to Senecio (Compositae). U.S. Nat. Herb. 34: 365–447, 1968.

220. Turner, B. L. and Morris, Michael I. New Taxa of Palafoxia (Astereae:Helenieae). Madrono 23: 79–80, 1975.

221. Keil, David J. A. Revision of Pectis Section Pectothrix (Compositae:Tageteae). Rhod. 79: 32–78, 1977.

222. Ibid. New Taxa in Pectis (Compositae:Pectidinae) from Mexico and the Southwestern United States. Brittonia 26: 30–36, 1974.

223. Powell, A. Michael. Taxonomy of Perityle Section Perityle (Compositae-Peritylinae). Rhod. 76: 229, 306, 1974.

224. Todson, T. K. A New Species of Perityle Compositae from Arizona. Journ. Ariz. Acad. 9: 35, 1974.

225. Powell, A. Michael. Taxonomy of Perityle Section Laphamia (Compositae-Helenieae-Peritylinae). Sida 5: 61–128, 1973.

226. Shinners, L. H. Species of Laphamia Transferred to Perityle (Compositae-Helenieae). Southw. Nat. 4: 204–209, 1959.

227. Anderson, L. R. Studies in Petradoria. Trans. Kan. Acad. Sci. 66: 632–684, 1964.

228. Ellison, William L. Taxonomy of Platyschkuhria (Compositae). Brittonia 23: 269–279, 1971.

229. King, R. M. Studies in the Eupatorieae (Compositae) 1–3, Rhod. 69: 35–42, 1967.

230. Johnson, R. Roy. Monograph of the Plant Genus Porophyllum (Compositae:Helenieae). Univ. Kan. Sci. Bull. 48: 225–267, 1969.

231. Tomb, A. Spencer. Reestablishment of the Genus Prenanthella Rydb. (Compositae-Cichorieae). Brittonia 24: 226, 1972.

232. Strother, John L. and Pilz, George. Taxonomy of Psathyrotes (Compositae:Senecioneae). Madrono 23: 24–40, 1975.

233. Ediger, Robert I. Revision of Section Suffruticosi of the Genus Senecio (Compositae). Sida 3: 504–524, 1970.

234. Cronquist, A. et al. Vascular Plants of the Pacific Northwest. Part 5: 288, 1955. Univ. Wash. Press, Seattle, Wash.

235. Barkley, T. M. Taxonomy of Senecio multilobatus and its Allies. Brittonia 20: 267–284, 1968.

236. Ibid. Intergradation of Senecio Sections Aurei, Tomentosi, and Lobati Through Senecio mutabilis Greenm. Southw. Nat. 13: 109–115, 1968.

237. Shinners, Lloyd H. Tessaria sericea (Nuttall) Shinners, Comb. Nov. (Compositae). Sida 3: 122, 1967.

238. Brown, R. C. and Schaack, Clark G. Two New Species of Tragopogon for Arizona. Madrono 21: 304, 1972.

239. Watson, Thomas J. Jr. The Taxonomy of Xylorhiza. Brittonia 29: 199–216, 1977.

240. Torres, Andres G. Taxonomy of Zinnia. Ibid. 15: 1–25, 1963.

INDEX

193

194

196